Industrial Ecology and Industry Symbiosis for Environmental Sustainability

Xiaohong Li

Industrial Ecology and Industry Symbiosis for Environmental Sustainability

Definitions, Frameworks and Applications

Xiaohong Li
Sheffield Hallam University
Sheffield, UK

ISBN 978-3-319-67500-8   ISBN 978-3-319-67501-5 (eBook)
https://doi.org/10.1007/978-3-319-67501-5

Library of Congress Control Number: 2017955202

© The Editor(s) (if applicable) and The Author(s) 2018
This work is subject to copyright. All rights are solely and exclusively licensed by the Publisher, whether the whole or part of the material is concerned, specifically the rights of translation, reprinting, reuse of illustrations, recitation, broadcasting, reproduction on microfilms or in any other physical way, and transmission or information storage and retrieval, electronic adaptation, computer software, or by similar or dissimilar methodology now known or hereafter developed.
The use of general descriptive names, registered names, trademarks, service marks, etc. in this publication does not imply, even in the absence of a specific statement, that such names are exempt from the relevant protective laws and regulations and therefore free for general use.
The publisher, the authors and the editors are safe to assume that the advice and information in this book are believed to be true and accurate at the date of publication. Neither the publisher nor the authors or the editors give a warranty, express or implied, with respect to the material contained herein or for any errors or omissions that may have been made. The publisher remains neutral with regard to jurisdictional claims in published maps and institutional affiliations.

Cover illustration: Pattern adapted from an Indian cotton print produced in the 19th century

Printed on acid-free paper

This Palgrave Pivot imprint is published by Springer Nature
The registered company is Springer International Publishing AG
The registered company address is: Gewerbestrasse 11, 6330 Cham, Switzerland

Preface

Why would I like to introduce and explore Industrial Ecology (IE) and Industrial Symbiosis (IS) from an operations management (OM) perspective?

IE considers developing and transforming industrial systems to industrial ecosystems, which have high levels of nearly closed-loop material exchanges and efficiency of energy cascading. IS, as one area within IE, explores ways of developing knowledge webs to facilitate the establishment of novel material exchanges and energy cascading within and across companies and industries. Applying IE and IS effectively dramatically improve environmental sustainability.

OM aims to continuously improve the efficiency and effectiveness of processes for producing goods and providing services to customers. However, a process in OM is considered as a linear transformation. Where raw materials come from and where industrial waste and after-use products go have not been considered purposefully. Linear transformation thinking needs to be replaced by closed-loop thinking in OM in order to successfully transform industrial systems to industrial ecosystems and be sustainable in the long run.

Industrial waste, by-products or residues generate environmental benefits, economic values and reduce the burden to societies if they can be reused, remanufactured or recycled as feedstock to another process. Reduction of disposals along all stages of a product life cycle, including the product after-use stage, contributes to environmental sustainability as well as economic and social sustainability. After-use products should also be considered for their re-entry to industrial systems through the application of reuse, remanufacturing and recycling options, with the assistance of design for environment (DfE). This allows after-use products to re-enter circulations in our industrial systems as resources instead of being disposed of into our natural environment. This can dramatically reduce the intake of virgin materials and pollutant and waste emissions to our natural environment.

This idea is not new and was proposed by Frosch and Gallopoulos in 1989. However, why are OM transformation processes still being considered and taught in a linear transformation representation of inputs and outputs? Shouldn't all processes and systems work towards nearly closed-loop industrial ecosystems in an extended system view?

It has not been an easy journey for me to fully understand the principles and purposes of IE and IS through exploring their definitions and applications in the literature, as the literature offers different and in some cases conflicting views. Writing this book has given me an opportunity to explore IE and IS in great depth and has clarified a number of queries I have had over the last few years. I hope this book can also help readers reappraise and elucidate their own thinking regarding IE and IS and their roles in achieving environmental sustainability. If that is the case, I have achieved my purpose.

Sheffield, UK,   Xiaohong Li
September 2017

Acknowledgements

I would like to thank Mr. Michael Leigh, my co-author for a journal article on Industrial Ecology and Industrial Symbiosis published in *the Journal of Cleaner Production* in 2015. I thank him for bringing my attention to Industrial Ecology and Industrial Symbiosis and the opportunities they presented for research within the Operations Management field. I would like to thank the editor Josie Taylor at Palgrave, whom I met at the European Operations Management conference (EurOMA) in 2015 for her great support for me to initialise this book proposal in 2016. I also thank the editorial assistant Lucy Kidwell and the editorial team in the Business and Management Scholarly & Reference Division, for their time and effort in communicating with me while I was working on this book. It has been challenging time wise to work on this book without being able to dedicate considerable time, energy and effort in a short space of time to bring this project to a conclusion sooner as planned, as I have a full time lecturing job at the Sheffield Business School, Sheffield Hallam University. Thanks to Lucy for her patience in bearing with me until the final date of submission.

I would like to give my huge thanks to my dear friends Dr. Lisa Trewhitt, Mrs. Nicola Livesey and Mr. Allan Dixon who had proofread my draft chapters. Without their proofreading, the book would not read as well as it does!

Thanks to my dear mother, Mrs. Xiangyun Yu, and my late father, Mr. Benling Li, who were both professors in Biology before their retirement, for their lifelong encouragement to my learning and achievement.

Thanks to my darling eight-year old daughter, little Miss Jasmine Dixon, for understanding that Mummy had to spend many weekends writing this book without being able to go out visiting places with her and her Daddy.

Finally, I wish to thank my readers in advance who are interested in making this world a better place by taking Industrial Ecology and Industrial Symbiosis into their learning and practice. It is you who will make the difference!

Contents

1 **An Introduction to Closed-Loop Concept in Industrial Ecology**   1
    References   8

2 **Industrial Ecology and Industrial Symbiosis - Definitions and Development Histories**   9
    2.1 Introduction   10
    2.2 Definitions of Industrial Ecology and Industrial Symbiosis   11
    2.3 Development Histories of Industrial Ecology and Industrial Symbiosis   25
    2.4 The Relationship between Industrial Ecology and Industrial Symbiosis and Study Areas within Industrial Ecology   31
    2.5 Summary   34
    References   36

3 Industrial Ecology Applications in the Four Areas — 39
3.1 Introduction — 40
3.2 Determining the Boundary of an Industrial Ecosystem — 42
3.3 Developing Symbiotic Relationships — 47
3.4 Improving Industrial Metabolism — 50
3.5 Aligning Legislation and Regulations to Support Industrial Ecology Applications — 53
3.6 Summary — 55
References — 58

4 Applications of Industrial Symbiosis — 61
4.1 Introduction — 62
4.2 Regional Community-based Industrial Symbiosis Development - Kalundborg in Denmark — 63
4.3 UK National Industrial Symbiosis Programme — 69
4.4 Eco-industrial Parks — 73
4.5 Conditions for IS Applications — 80
4.6 Summary — 85
References — 86

5 Life Cycle Thinking and Analysis, Design for Environment, and Industrial Ecology Frameworks — 91
5.1 Introduction — 92
5.2 Life Cycle Thinking/Analysis — 92
5.3 Design for Environment — 96
5.4 Industrial Ecology Conceptual Frameworks — 102
5.5 Summary — 106
References — 107

6	**Challenges for Applying Industrial Ecology and Future Development of Industrial Ecology**	**111**
	6.1 Introduction	112
	6.2 Challenges for Applying Industrial Ecology and Industrial Symbiosis	112
	6.3 Future Development of Industrial Ecology	121
	6.4 Summary	121
	References	124
Index		**127**

Abbreviations

BCSD-UK	Business Council for Sustainability Development-United Kingdom
C2C	Cradle to cradle
C2G	Cradle to grave
Defra	Department for Environment, Food and Rural Affairs
DETDZ	Dalian Economic and Technological Development Zone (Dalian, China)
DfE	Design for environment
DfRec	Design for recycling
DfRem	Design for remanufacture
DSP	Dominant Social Paradigm
EIC	Eco-industrial cluster
EIP	Eco-industrial park
EPA	Environmental Protection Agency (USA)
EPR	Extended producer responsibility
IE	Industrial Ecology
IM	Industrial metabolism
IP	Industrial park
IS	Industrial Symbiosis
JIT	Just-in-time
LCA	Life cycle assessment

LCC/LCCA	Life cycle cost analysis
LCEcA	Life cycle economic analysis
LCEnA	Life cycle environmental analysis
LCI	Life cycle inventory
LCIA	Life cycle impact assessment
LCRA	Life cycle risk analysis
LCSA	Life cycle sustainability analysis
LCSoA	Life cycle social analysis
MEP	Ministry of Environmental Protection (MEP)
NISP	National Industrial Symbiosis programme (UK)
OEM	Original equipment manufacturer
OM	Operations management
SEPA	State Environmental Protection Agency (SEPA) (China)
SLCA	Social life cycle analysis
TEDA	Tianjin Economic-Technological Development Area (Tianjin, China)
TQM	Total Quality Management
WEEE	Waste Electrical and Electronic Equipment

List of Figures

Fig. 1.1	A linear transformation process	3
Fig. 1.2	A closed-loop OM process thinking	4
Fig. 2.1	Development history of Industrial ecology	26
Fig. 2.2	Development history of Industrial symbiosis	28
Fig. 2.3	Industrial ecology and its interrelated four study areas	32
Fig. 4.1	Some solid waste symbiotic exchanges and external material inputs and outputs excluding water and energy of some businesses in Kalundborg	68
Fig. 5.1	Product life cycle with consideration of product after-use options	93
Fig. 5.2	Design for after-use options and contribution of DfE to closed-loop industrial ecosystems	101
Fig. 5.3	A conceptual IE framework at a factory level	103
Fig. 5.4	A conceptual IE framework at a supply chain level	105
Fig. 6.1	The fundamental role of IE to environmental sustainability, and further to social and economic sustainability	114

List of Tables

Table 2.1	Industrial ecology (IE) definitions, associated key features and comments	12
Table 2.2	Industrial symbiosis (IS) definitions, associated key features and comments	20
Table 3.1	Key features of approaches for determining industrial ecosystem boundary	43
Table 3.2	Advantages and disadvantages of approaches for determining the industrial ecosystem boundary	44
Table 3.3	Key aspects for applications in each of the four IE areas	56
Table 4.1	Typical novel exchanges of symbiotic collaborations within Kalundborg	65
Table 4.2	Characteristics of EIPs in the USA, the Netherlands, China and South Korea	74
Table 5.1	Life cycle analysis tools in the literature and some suggested terms and abbreviations	94
Table 5.2	Design for environment (DfE) and associated approaches and strategies	97

Table 5.3	Upgraded end-of-life pollution control approaches after integrating with 'design for after-use' pollution prevention approaches in DfE	99
Table 6.1	Future development elements of IE in four areas for research and applications	122

1
An Introduction to Closed-Loop Concept in Industrial Ecology

Abstract This introductory chapter explains the fundamental problem of a linear transformation representation used in operations management (OM) to the development of environmental sustainability. Linear transformation thinking needs to be replaced by closed-loop system thinking. Industrial Ecology can help achieve this development. This chapter explores the basic concepts in relation to IE, including biological ecosystem, industrial ecosystem, sub-ecosystems and their interactions with the ecosystem of the Earth. IE considers the development of high-level closed-loop industrial ecosystems as its ultimate goal through mimicking key principles of biological ecosystems. An industrial ecosystem needs to work towards high-level closed-loop material exchanges and high efficiency of energy cascading. The system boundary is subject to study purposes and extended system thinking should be applied.

Keywords Linear transformation · Closed-loop material exchanges Biological ecosystem · Industrial ecosystem · Industrial Ecology Industrial Symbiosis

Environmental sustainability has become increasingly important as it has a fundamental role to achieve economic and social sustainability (Goodland 1995; Goodland and Daly 1996). This book focuses on Industrial Ecology (IE) as a study field and one of its areas—Industrial Symbiosis (IS) and how IE and IS can contribute to environmental sustainability, by promoting closed-loop representations of processes, particularly from an operations management (OM) perspective.

Humans have made an impact on the Earth. We are prosperous and are producing plenty, more than we need. The productivity of the industrial world is increasing constantly and we produce more and faster. We use more natural resources and dispose of more waste to the environment. We are facing natural resource shortages and a high level of pollution and waste. We have been searching for ways to mitigate the situation but much greater effort from all study disciplines and practices is still needed. OM is concerned with producing goods and providing services and improving processes of industrial design, production and delivery. OM also needs to contribute to environmental sustainability. This book explores IE and IS from the OM perspective, but the contents are relevant to all other disciplines in both managerial and technical subject fields, as IE is an interdisciplinary study field and IS supports the achievement of IE's ultimate goal of developing industrial ecosystems for achieving environmental sustainability.

Currently, OM considers a process in a linear transformation and aims to improve the efficiency and effectiveness of this linear transformation process. OM has not purposefully considered where and what types of resources are coming from, and where waste or end-life products should go. Some OM researchers have been exploring environmental sustainability issues in relation to industrial activities. However, environmental sustainability has not been fully embedded in OM, particularly when the process analysis in OM still relies on a linear transformation representation.

In a linear transformation process, staff and facilities provide the capacity to transform raw materials and energy as inputs into finished goods as well as waste or by-products as outputs (Fig. 1.1). Waste or by-products are normally not considered in the analysis of this linear transformation process. That is what is still taught in OM. As staff and

1 An Introduction to Closed-Loop Concept in Industrial Ecology

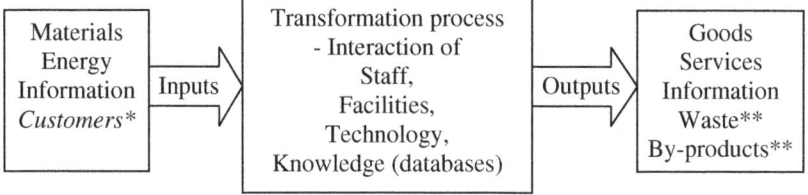

* For service industry only
** Also termed industrial residues

Fig. 1.1 A linear transformation process

facilities interacting with each other to provide the capacity of a process, OM focuses on the improvement of staff and facilities and their interactions to be more efficient and effective. However, OM has not paid much attention to industrial waste generated, except for reducing waste along industrial processes to be more efficient. OM focuses on providing goods and services and has not considered product after-use options, which can also add economic values to a company and its supply chain, as well as environmental values. OM aims to achieve operations objectives of cost, quality, speed, dependability and flexibility, but neglects sustainability and gives limited consideration to sustainability performance, including environmental sustainability. It seemed acceptable when natural resources were plentiful and waste generated was far less than the Earth's carrying capacity. However, this is no longer the case. Industrial systems cannot be sustained for long when natural materials become scarce and waste to nature is more than its carrying capacity—the ability with which nature can decompose waste within a given time and space.

For industrial systems to be sustainable and also the natural world where we are living sustainable, industrial activities need to contribute to the sustainability of the natural world. The linear transformation thinking needs to be replaced by the closed-loop system thinking. The idea of closed-loop industrial (eco)systems is not new. Nearly thirty years ago, Frosch and Gallopoulos (1989) proposed the industrial

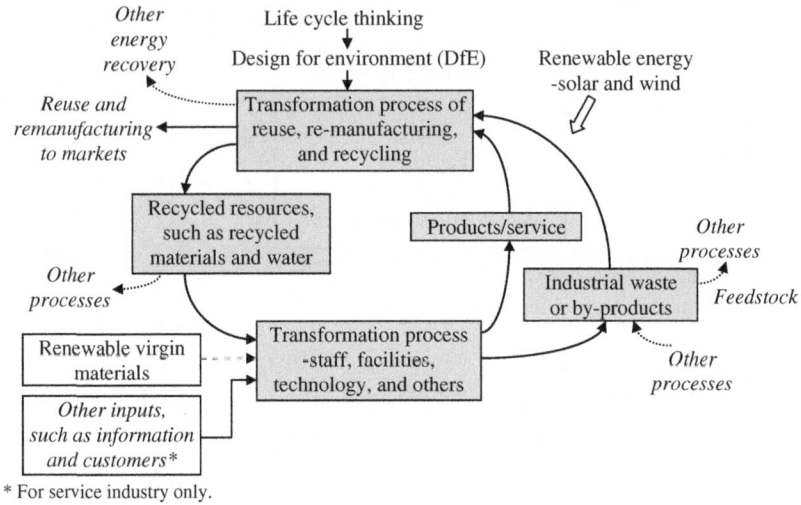

Fig. 1.2 A closed-loop OM process thinking

ecosystem concept, based on the closed-loop system thinking under which industrial systems mimic the key principles of biological ecosystems. The waste from one process can be the feedstock to another (Frosch and Gallopoulos 1989). Achieving this on a large scale requires novel ideas in order for different industries to work together to reduce waste generation as well as increase reuse, remanufacturing and recycling of industrial waste. Reuse, remanufacturing and recycling of after-use products also need to be dramatically increased, with the aid of design for environment (DfE). Recycled materials used in industrial processes still represent a fraction of the total materials used and using the waste from one process as the feedstock to another remains limited. Reuse, remanufacturing and recycling of after-use products are far from achieving their full potential. The linear transformation process is still considered in OM and much more needs to be done. The closed-loop system thinking should be adopted in OM analyses to support business decision-making (Fig. 1.2).

As a starting point, materials can be categorised as recycled materials—coming from industries or consumers through recycling companies, or renewable virgin materials—coming from nature systems, which are not in shortage in supply and are renewable over an acceptable time period. The proportion of recycled materials over the total materials taken into the circulations of an industrial system needs to be constantly increased towards 100 per cent. We need innovative ideas in order for waste or by-products from one process to be the feedstock to another. We also need innovative designs of products and processes to allow industrial waste and after-use products to be reused, remanufactured or recycled. Management and OM specialists, environmental scientists and material engineers need to work together to propose innovative ideas on reuse, remanufacturing and recycling, as well as using feedstock to other processes.

The closed-loop system thinking is rooted in IE (Frosch and Gallopoulos 1989; Tibbs 1992), which offers approaches to develop industrial ecosystems which have high level of closed-loop material exchanges and high efficiency of energy cascading. IE and IS do not view waste as waste, but resources including energy (Frosch and Gallopoulos 1989; Malcolm and Clift 2002; Tibbs 1992). In this introductory chapter, fundamental concepts in relation to IE and IS are explained leading to the closed-loop system thinking. These concepts include biological, ecological, system and ecosystem. The meaningful combinations of these four concepts with which IE and IS are concerned are biological system, biological ecosystem, industrial system and industrial ecosystem. In order to appreciate IE and IS, these fundamental concepts need to be explored through addressing the following questions.

1. Are all biological systems biological ecosystems?
2. Is there a boundary for a biological ecosystem? If yes, what determines the boundary of a biological ecosystem?
3. What qualifies an industrial system as an industrial ecosystem?

An ecosystem commonly refers to a biological ecosystem, which includes abiotic components which are non-living and environmental elements, such as sunlight, temperature, precipitation, water and moisture, soil and water chemistry etc., as well as biotic components, which are living elements, like plants, animals and humans, and bacteria and fungi (Borman and Liken 1970; Wessells and Hopson 1988; University of Michigan 2008-online teaching materials). These components interact with each other and transform their forms over time and space in an ecosystem or across different ecosystems. A transformation takes place in a material exchange and energy cascading process. The key features of biological ecosystems are:

1. Materials/elements cycle in nearlyclosed-loop system in subsystems to the Earth and the ecosystem of the Earth cycles its materials in a closed-loop system over time and space, and
2. Energy flows in an open system within a subsystem to the Earth and within the ecosystem of the Earth, with the input of solar energy from the sun (Wessells and Hopson 1988).

An ecosystem's boundary is not naturally fixed and it is only a choice for study purposes, as we can study a fish pond or we can study the Earth as a whole. Each of the sub-ecosystems to the ecosystem of the Earth cannot be a 100 per cent closed-loop system in terms of its material exchanges as it interacts and exchanges materials with other sub-ecosystems and to the ecosystem of the Earth to some degree. When a biological ecosystem, a sub-ecosystem to the ecosystem of the Earth, maintains its balance in which waste generated within the ecosystem does not exceed its own carrying capacity, the ecosystem sustains in the space and time specified. Many biological ecosystems interact with each other to form the ecosystem of the Earth. When waste generated by the overall ecosystem does not exceed the carrying capacity of the Earth, the ecosystem of the Earth sustains. The position of a chosen ecosystem for a study in relation to the ecosystem of the Earth and its interaction with other ecosystems need to be considered in a study whenever possible. However, for the feasibility of a study, a selected ecosystem with a defined boundary is used for exploration.

We are certain that not all industrial systems are industrial ecosystems and many are far from being industrial ecosystems. An industrial ecosystem should cycle its materials in a nearly closed-loop manner, like a biological ecosystem. We have acknowledged that even biological sub-ecosystems, as we human beings define them for study purposes, within the ecosystem of the Earth do not have 100 per cent closed-loop material exchanges The question is: at what level of 'close-loop' material exchanges, does an industrial system qualify itself as an industrial ecosystem? We cannot subjectively propose a threshold level above which an industrial system can be considered an industrial ecosystem. However, the relativity of the closed-loop material exchanges from a lower level to a higher level leads to a meaningful measure for improved environmental sustainability.

Designing and developing industrial systems towards industrial ecosystems have a profound and fundamental long-term positive impact on the overall environmental sustainability development of the Earth. As no ecosystems except for the ecosystem of the Earth can ever qualify 100 per cent closed-loop material exchanges, the measure for an industrial system to be qualified as an industrial ecosystem is its continuous movement towards higher levels of closed-loop material exchanges. The industrial revolution has increased the rate of production dramatically, which exaggerates the defect of the linear transformation process on which the traditional industrial systems are based. The idea of industrial ecosystems breaks the traditional view of the linear transformation process. Instead, it considers a process or system in a closed-loop manner of material exchanges based on extended system thinking.

Following this introductory chapter, definitions and development histories of IE and IS and their relationship are presented in Chap. 2. Chapter 3 explores applied elements in IE's four interrelated areas: industrial ecosystem, IS, industrial metabolism (IM) and legislation and regulations for IE development and applications. Chapter 4 reviews IS applications in three forms: a regional IS practice, a national IS programme and eco-industrial parks (EIPs), and also explores conditions for IS applications. Chapter 5 presents life cycle thinking, DfE and frameworks for achieving integration and collaboration for synergies to reduce the degree of interaction with the natural environment. Chapter 6 explores challenges in applying IE and IS and future development in IE's four areas.

References

Borman, F. H., & Likens, G. E. (1970). The nutrient cycles of an ecosystem. *Scientific American, 223*(October), 92–101.
Frosch, R. A., & Gallopoulos, N. E. (1989). Strategies for manufacturing. *Scientific American, 261*(September), 144–152.
Goodland, R. (1995). The concept of environmental sustainability. *Annual Review of Ecology and Systematics, 26,* 1–24.
Goodland, R., & Daly, H. (1996). Environmental sustainability: Universal and non-negotiable. *Ecological Applications, 6,* 1002–1017.
Malcolm, R., & Clift, R. (2002). Barriers to Industrial Ecology, The strange case of "The Tombesi Bypass". *Journal of Industrial Ecology, 6,* 4–7.
The University of Michigan. (2008). *The concept of the ecosystem.* https://globalchange.umich.edu/globalchange1/current/lectures/kling/ecosystem/ecosystem.html. Accessed 02 Sept 2016.
Tibbs, H. (1992). *Industrial ecology, an environmental agenda for industry* (pp. 4–19). Winter: Whole Earth Review.
Wessells, N. K., & Hopson, J. L. (1988). *Biology, Chapter 44.* New York: Random House.

2
Industrial Ecology and Industrial Symbiosis - Definitions and Development Histories

Abstract Various definitions of Industrial Ecology (IE) and Industrial Symbiosis (IS) have been provided in the literature over the past thirty years. These definitions have offered some insights but also confusion due to inconsistency. IE, as an interdisciplinary study field, develops and applies different approaches in its four interrelated areas: industrial ecosystem, IS, industrial metabolism (IM) and environmental legislation and regulations. The ultimate goal of IE is to develop nearly closed-loop industrial ecosystems to enhance environmental sustainability. IS focuses on the development of knowledge webs of novel material, energy and waste exchanges to facilitate the establishment of synergies to support the achievement of this IE goal. The difference between IE and IS lies in the focus, instead of the scale of economy.

Keywords Industrial Ecology · Industrial Eymbiosis · Definitions Development histories · Relationships between Industrial Ecology and Industrial Symbiosis

2.1 Introduction

A number of researchers have provided definitions and explored applications of Industrial Ecology (IE) and Industrial Symbiosis (IS) (e.g. Chertow 2000; Despeisse et al. 2012; Ehrenfeld 1997; Heeres et al. 2004; Lombardi and Laybourn 2012; Lowe and Evans 1995; Mirata 2004; Park et al. 2016; Tian et al. 2014; Tibbs 1992; Valentine 2016). The proposed definitions have clarified the important roles of IE and IS in achieving environmental sustainability to some extent and have inspired research and applications of IE and IS. However, definitions of IE and IS offered in the literature are not always consistent. Some definitions distinguish IE and IS; whilst others do not. Some definitions set conditions for IS applications, such as geographic proximity and diverse industries (Chertow 2000; Boix et al. 2015); whilst others considered these conditions unnecessary (Branson 2016; Lombardi and Laybourn 2012; Jensen et al. 2011). Differences in understanding IE and IS can stimulate debates. However, unclear concepts can cause confusion and do not support IE and IS development and applications. Therefore, clarifying IE and IS concepts through exploring existing definitions is necessary. Comparing key features of IE and IS defined in the literature helps generate a more consistent understanding of IE and IS which assists their future development and applications.

IS certainly cannot exist alone without the presence of IE, as IS supports the implementation of IE principles and the achievement of the IE goal. However, it is not always the case that reported IS applications have considered IE principles, particularly for a number of eco-industrial parks (EIPs) applications (Gibbs and Deutz 2005). In addition, there are different understandings of the relationship between IE and IS in the literature. Hence, reviewing the development histories of these two concepts and critically evaluating the relationship between IE and IS presented in the literature can help clarify some confusion.

Hence, this chapter explores key definitions of IE and IS and their development histories. The common themes of IE and IS and divergences in the interpretation of these two concepts in the literature are explored and compared, along with their relationships.

2.2 Definitions of Industrial Ecology and Industrial Symbiosis

Key definitions of IE in the literature are presented in Table 2.1 with identified key features of each definition followed by comments. Identifying and extracting a definition of IE is not always straightforward because explanations of IE and related aspects are fairly often provided within the content of a research paper, rather than being presented in a standard definition format.

Some early definitions referred to IE as a method, an approach or a framework. The definitions have gradually established that IE is an interdisciplinary study field. Like its counterpart biological ecology, IE is a study field which contains different methods, approaches and frameworks to design and transform industrial systems to nearly closed-loop industrial ecosystems. Considering IE only as an approach, a framework or a method restricts its development and applications.

These definitions also propose other related concepts, such as carrying capacity, sub-industrial ecosystems and mother ecosystems. Biological sub-ecosystems and industrial sub-ecosystems are subject to their own carrying capacities as well as the carrying capacity of the ecosystem of the Earth, which is a closed-loop system of material exchanges, but an open system for energy flow. Not all the definitions focus on or reveal the core of the IE, which is the design and transformation (development) of industrial systems to nearly closed-loop industrial ecosystems.

IE development has been built upon the understanding of IE related concepts/areas, such as industrial ecosystem, IS and industrial metabolism (IM) (Tibbs 1992). The term of IS has existed for a long time and many early publications of IE mentioned IS (Ehrenfeld and Gertler 1997; Lowe and Evans 1995; Tibbs 1992). However, definitions of IS were very rarely provided prior to 2000, except for one study by Ehrenfeld and Gertler (1997) which defined and explained IS using Kalundborg as the case to explore IE in practice. The intensity of studying IS began early this century. Chertow offered the most quoted definition of IS in 2000. Since 2000, there have been an increased number

Table 2.1 Industrial ecology (IE) definitions, associated key features and comments (*Note* 'Underline' for Industrial ecology or IE is added by the author)

Definitions of Industrial Ecology (IE)	Key features	Comments
'The industrial ecosystem would function as an analogue of biological ecosystem' (Frosch and Gallopoulos 1989, p. 144).	Industrial ecosystems mimic biological ecosystems.	This entry does not define IE directly but describes the core of IE, which is to develop industrial ecosystems of closed-loop or nearly closed-loop material and energy exchanges through integrating industrial processes.
'The traditional model of industrial activity… should be transformed into a more integrated model: an industrial ecosystem. In such a system the consumption of energy and materials is optimised, waste generation is minimised and the effluents of the one process … serve as raw material for another process' (Frosch and Gallopoulos 1989, p. 144).	An industrial ecosystem is an integrated model of industrial activities. Industrial ecosystems optimise the consumption of energy and materials and minimise waste through linkages between industrial processes.	This entry does not indicate that an incustrial ecosystem is part of the natural system.
'Equally important is the way in which the inputs and outputs of individual processes are linked within the overall industrial ecosystem. This linkage is crucial for building a closed or nearly closed system' (Frosch and Gallopoulos 1989, p. 149).	An industrial ecosystem is a closed-loop or nearly closed-loop system.	This entry does not imply the need for crossing industrial boundaries for IE. This proposes two questions: what is the overall industrial ecosystem and what is its boundary?

(continued)

Table 2.1 (continued)

Definitions of Industrial Ecology (IE)	Key features	Comments
'Industrial ecology involves designing industrial infrastructure as if they were a series of interlocking manmade ecosystems interfacing with the natural global ecosystem' (Tibbs 1992, p. 5).	IE is to design interlocked industrial ecosystems which interact with the ecosystem of the Earth.	Adds value by relating IE to an extended system view of industrial ecosystems, their relationships with the natural global ecosystem, improved efficiency of industrial systems, and the concept of the carrying capacity of the natural ecosystem
'The aim of industrial ecology is to interpret and adapt an understanding of the natural system and apply it to the design of the manmade system, in order to achieve a pattern of industrialisation that is not only more efficient, but that is intrinsically adjusted to the tolerances and characteristics of the natural system' (Tibbs 1992, p. 6).	IE is to understand the natural system. IE is to apply an understanding of natural systems to design manmade industrial systems. IE considers both the efficiency of the industrial system and the carrying capacity of the natural global ecosystem within which it is placed.	The last part in this entry indicates that IE is an interdisciplinary study field, including both management disciplines (such as operations management and environment management) and technical (or technology and design) aspects related disciplines (such as environmental science and material engineering).
'Industrial ecology permits an integrated managerial and technological interpretation' (Tibbs 1992, p. 8).	IE contains both managerial and technical aspects.	
'Industrial ecology is a new approach to the industrial design of products and processes and the implementation of sustainable manufacturing strategies' (Jelinski et al. 1992, p. 793).	IE is a new approach for sustainable manufacturing.	IE is not just an approach but a study of many approaches.
'Industrial ecology seeks to optimise the total materials cycle from virgin material to finished materials, to component, to product, and to ultimate disposal' (Jelinski et al. 1992, p. 793).	IE is to optimise the total materials cycle.	IE actually goes beyond the optimisation of the total materials cycle as described in this entry, from virgin material to ultimate disposal as a linear process. This definition could mislead by neglecting the development of nearly closed-loop industrial ecosystems by IE.

(continued)

Table 2.1 (continued)

Definitions of Industrial Ecology (IE)	Key features	Comments
'Industrial ecology is an emerging framework for environmental management, seeking transformation of the industrial system in order to match its inputs and outputs to planetary and local carrying capacity. A central IE goal is to move from a linear to a closed-loop system in all realms of human production and consumption' (Lowe and Evans 1995, p. 47).	IE is a framework. IE is for environmental management. IE is to transform industrial systems to industrial ecosystems. Industrial ecosystems need to match their inputs and outputs to the carrying capacity of local biological ecosystems and the ecosystem of the Earth.	IE is more than a framework, but a study of a number of frameworks. IE is an interdisciplinary study field, not just for environmental management. Considers both planetary (global) and local carrying capacity for industrial ecosystems to match in design and transformation.
'Industrial ecology offers a theoretical foundation to support the transformation to a sustainable industrial system, operating in this balanced fashion (production and decomposition are well balanced, with nutrients recycling continuously to support the next cycles of production)' (Lowe and Evans 1995, p. 48).	IE aims to move a linear to a closed-loop industrial system. IE offers a theoretical foundation. An industrial ecosystem operates in the balance of production and decomposition.	Emphasises the transformation from a linear to a (nearly) closed-loop system as the core of IE. Highlights the importance of the balance between production and decomposition in closed-loop industrial ecosystems.

(continued)

Table 2.1 (continued)

Definitions of Industrial Ecology (IE)	Key features	Comments
'Industrial ecology is the means by which humanity can deliberately and rationally approach and maintain sustainability, given continued economic, cultural, and technological evolution. The concept requires that an industrial system be viewed not in isolation from its surrounding systems, but in concert with them. It is a systems view in which one seeks to optimise the total materials cycle from virgin material to finished material, to component, to product, to obsolete product, and to ultimate disposal. Factors to be optimised include resources, energy and capital' (Graedel and Allenby 1995/2003, p. 18).	IE is a means to deliberately and rationally approach and maintain sustainability. IE views industrial systems not in isolation from its surrounding systems, but in concert with them. IE is to optimise the total materials cycle, including resources, energy and capital. IE is related to global sustainability.	IE is not just a means, but a study field. Specifies the element of 'consciousness' in relation to IE for improving environmental sustainability. 'Capital' is associated with 'resources' or 'energy' in monetary value. Hence it should not be individually listed. IE is beyond the optimisation of the total materials cycle as described in this entry, from virgin material to ultimate disposal as a linear process.
'Industrial ecology is the study of technological organisms, their use of resources, their potential environmental impacts, and the ways in which their interactions with the natural world could be restructured to enable global sustainability' (Graedel and Allenby 1995/2003, p. 39).		This definition could mislead by neglecting the development of nearly closed-loop industrial ecosystems by IE.

(continued)

Table 2.1 (continued)

Definitions of Industrial Ecology (IE)	Key features	Comments
'Industrial ecology goes further. The idea is first to understand how the industrial system works, how it is regulated, and its interaction with the biosphere; then on the basis of what we know about ecosystems, to determine how it could be restructured to make it compatible with the way natural ecosystems function' (Erkman 1997, p. 1).	IE is to understand the industrial system and the natural system and their interaction. IE is to restructure the industrial system for it to be compatible with the natural ecosystem.	Emphasises the 'compatibility' of industrial systems with natural systems. How IE can restructure industrial systems to be compatible with natural ecosystems remains unexplored.
'Industrial ecology, in its paradigmatic form, would become part of a new evolving Dominant Social Paradigm (DSP) that would include the maintenance of the natural world as a fundamental normative goal' (Ehrenfeld 1997, p. 88).	IE is part of a new Dominant Social Paradigm (DSP). IE is to maintain the natural world.	Adds value by setting IE within a wider context by considering IE as part of a new DSP. Emphasises the fundamental role of IE in achieving environmental sustainability in a very general way.
'Industrial ecology symbiotically links industries so that environmental conscious practices can also be profitable. To do this, industrial ecology uses principles of biological ecosystems to optimise the flows and transformation of materials and energy within and across the boundaries of industrial systems' (Dunn and Steinemann 1998, p. 661).	IE is to establish symbiotic relationships across industries. IE can also lead economic gains. IE uses principles of biological ecosystems. IE is to optimise the flows and transformations of materials and energy, within and across the boundaries of industrial systems.	Emphasises the importance of symbiotic relationships among industries in IE, (as now this is specifically considered as IS, which is one area within IE). Specifies that IE is for both within an industrial system boundary and also across different industrial system boundaries. The nearly closed-loop ecosystem development by IE is not specified.

(continued)

Table 2.1 (continued)

Definitions of Industrial Ecology (IE)	Key features	Comments
'In a perfect IE [sic. refers to an industrial ecosystem] both of the systems (the industrial (sub) system and the (mother) ecosystem) operate according to the same principles of system development: roundput, diversity, locality and gradual change' (Korhonen 2001, p. 257).	A perfect industrial ecosystem operates like its mother ecosystem. Industrial ecosystem follows: roundput (closed-loop), diversity, locality and gradual change.	Specifies the four principles of biological ecosystems for industrial ecosystems to mimic/follow. It is debatable for locality (geographic proximity) in the literature.
'IE should be defined as a field of study (or branch of science) concerned with the interrelationships of human industrial systems and their environments' (Seager and Theis 2002, p. 226).	IE is a field of study. IE is concerned with interrelationships of industrial systems and their environment.	Specifies that IE is a study field. This is a very general definition of IE. Therefore, it is unlikely to be used for IE applications. The closed-loop ecosystem development needs to be specified.
'… industrial ecology draws on some vision of an ecological network of interconnected actors exchanging matter and energy. Some see the metaphor as ontological–a way of extending the bounds of thinking; others see the metaphor as normative, providing prescriptive guides for designing a more sustainable world' (Ehrenfeld 2004, p. 827).	IE concerns an ecological network of interconnected actors exchanging matter and energy. IE can be a new way of thinking and/or a practical guide for designing a more sustainable world.	Emphasises the ecological requirement of IE in symbiotic relationships (IS).

(continued)

Table 2.1 (continued)

Definitions of Industrial Ecology (IE)	Key features	Comments
'Its (IE's) key feature lies in the integration of various components of a system to reduce its net resource input as well as pollutant and waste outputs' (Despeisse et al. 2012, p. 31).	IE is concerned with the integration of various components of a system. IE reduces net resource inputs and pollutant and waste outputs of a system.	Emphasises the integration of different components in a system in IE. Emphasises that the goal of IE is to reduce inputs and outputs of industrial systems and also implying the concept of closed-loop material exchanges.
'IE considers principles of biological ecosystems when designing and redesigning industrial systems to create more efficient interactions both within industrial systems and between industrial systems and natural systems' (Leigh and Li 2015).	IE considers principles of biological ecosystems. IE designs and redesigns industrial systems. IE improves the efficiency of interactions within and between industrial systems and the natural system.	Emphasises the consideration of principles of biological ecosystems in IE, but does not specify them. Emphasises interactions between systems, industrial or natural. The closed-loop ecosystem development needs to be specified.

of studies exploring IS and its applications. Chertow (2000) considered eco-industrial parks (EIP) as 'concrete realisation' of IS even though the applications of EIPs started early in 1990 in the USA, followed by other countries around the world. On the one hand, many studies of IS explained IE first and made direct relevance of IS in its relationship to IE (Chertow 2000; Costa and Ferrão 2010; Lombardi and Laybourn 2012). On the other hand, some applications of IS have neglected the relevance of IS to IE in terms of its ultimate goal of developing nearly closed-loop industrial ecosystems. Key definitions of IS and associated features followed by comments are presented in Table 2.2.

Early definitions of IS focused on efficiency and optimisation of resources or resource flows without emphasising the eco element or novelty of exchanges based on the establishment of symbiotic relationships. The definition by Domenech and Davies (2011) made a breakthrough, by stressing the need of a web of knowledge to facilitate the establishment of physical exchanges of resources among diverse organisations. This emphasises the importance of knowledge in IS development. The definition by Lombardi and Laybourn (2012) further clarifies that the exchanges must be novel.

These definitions clearly confirm that IS is part of IE. However, most definitions need to make it clearer that IS is to fulfil the goal of IE, which is to develop high levels of nearly closed-loop industrial ecosystems. These definitions also need to address the system boundary aspect.

IS requires the integration of the following features:

1. A web of knowledge,
2. A network of diverse organisations,
3. Novel sourcing of inputs,
4. Value-added destinations of non-product outputs (and further end-life products),
5. Improved business and technical processes, and
6. A collective approach of a system as a whole.

The question here is how to define the boundary of an industrial ecosystem, which is addressed in Chap. 3. The further understanding of IE and IS can be gained through exploring their development histories.

Table 2.2 Industrial symbiosis (IS) definitions, associated key features and comments
(*Note* 'Underline' for Industrial symbiosis or IS is added by the author)

Definitions	Key features	Comments
'Industrial symbiosis is closely related [with IE] and involves the creation of linkages between firms to raise the efficiency, measured at the scale of the system as a whole, of material and energy flows through the entire cluster of processes' (Ehrenfeld and Gertler 1997, p. 68).	IS is closely related to IE. IS involves the creation of linkages between firms. IS raises the efficiency of a system as a whole.	Clearly indicates the close relation between IE and IS and also IS focuses on the creation of linkages between firms. Emphasises the importance of the entire system. The linkages for creating novel material exchanges to increase the level of the closed-loop are not indicated. Mentioning that IS is for raising efficiency of material and energy flows can be misleading. The system boundary and how to determine the system boundary remain unaddressed.
'Industrial symbiosis, as part of the emerging field of industrial ecology, demands resolute attention to the flow of materials and energy through local and regional economies' (Chertow 2000, p. 313).	IS is part of IE which deals with the flow of materials and energy through local and regional economies, but not the global economy.	This is a well-cited definition. However, it raises a critical question regarding whether IS and IE should be distinguished by the scale of economy. IS is part of IE.
'Industrial symbiosis engages traditionally separated industries in a collective approach to competitive advantage involving physical exchange of materials, energy, water, and/or by-products. The keys to industrial symbiosis are collaboration and synergistic possibilities offered by geographic proximity' (Chertow 2000, p. 314).	IS engages traditionally separated industries. The keys to IS are collaboration and synergistic possibilities. IS is under the condition of geographic proximity.	Emphasises the importance of crossing industrial boundaries using a collective approach for IS. Adds value by considering different physical exchanges, not just waste and by-products for IS. This definition raises another critical question regarding whether geographic proximity is essential for IS. (There have been many symbiotic exchanges between companies across regions reported in the literature.) Another question is whether developing physical exchanges is the only concern in IS?

(continued)

Table 2.2 (continued)

Definitions	Key features	Comments
'As a sub-discipline of industrial ecology, industrial symbiosis is concerned with resource optimisation among co-located companies' (Jacobsen 2006, p. 239).	IS is a sub-discipline of IE. IS is concerned with collective resource optimisation among co-located companies.	IS is part of IE. This definition considered 'collective resource optimisation' and 'co-located firms' as key features of IS. This can be misleading. IS is to establish symbiotic relationships among different industrial firm. By doing so, IS contributes to collective resource optimisation to co-located facilities, but not just for co-located facilities alone. IS supports the achievement of the IE goal, which is to develop nearly closed-loop industrial eco-systems for improved environmental performance.
'Within this framework of inter-firm relationships, <u>industrial symbiosis</u> (IS) can be categorised as a concept of collective resource optimisation based on by-product exchanges and utility sharing among different co-located facilities' (Jacobsen 2006, p. 240).	IS is based on by-product exchanges and utility sharing.	IS is more than just by-product exchanges and utility sharing among different co-located facilities.
'Thus, at least three different entities must be involved in exchanging at least two different resources to be counted as a basic type of <u>industrial symbiosis</u>' (Chertow 2007, p. 12).	IS requires the involvement of three different entities. IS requires two different resource exchanges.	This 2–3 rule has absolutely no grounding. The number of entities and resources involved to qualify a basic type of IS given in this definition is totally subjective. It is not the involvement of the number of entities or resource, but the type of exchange that qualifies a basic type of IS. The type of exchange should be a novel exchange supporting the development of a higher level of closed-loop material exchanges and efficiency of energy cascading. Interestingly, a symbiotic relationship in the biological ecology involves only two species. (Two plus two becomes four and it will be more as long as we can have more twos in the right type of exchanges.)

(continued)

Table 2.2 (continued)

Definitions	Key features	Comments
'Within the field of industrial ecology, Industrial Symbiosis (IS) has emerged as a body of exchange structures to facilitate progress to a more eco-efficient industrial system. By establishing a collaborative web of knowledge, material and energy exchanges among different organisational units, IS networks aim to reduce the intake of virgin materials and lower the production of waste by the industrial sector' (Domenech and Davies 2011, p. 79).	IS is within the field of IE.	IS is part of IE.
	IS is a body of exchange structures to facilitate progress to a more eco-efficient industrial system.	IS focuses on the establishment of a network of collaborations (exchange structures) to achieve the development of a more eco-efficient industrial system. The question is whether eco-efficiency always supports the closed-loop development.
	IS establishes a collective web of knowledge, material and energy exchanges.	Emphasises the eco-efficiency which IS aims for, rather than just efficiency or optimisation of resources.
		Emphasises the establishment of knowledge webs in IS, leading to physical exchanges. This element is a breakthrough in IS definitions.
	IS networks aim to reduce the intake of virgin materials and lower the production waste.	Specifically mentions that IS networks aim to reduce virgin material intake and waste production outputs.
		Whether the system boundary is determined by an industrial sector is certainly debatable.
'Industrial symbiosis examines cooperative management of resource flows through networks of businesses known in the literature as industrial ecosystems' (Chertow and Ehrenfeld 2012, p. 13).	IS examines cooperative management of resource flows.	The definition is too general. It does not provide the focus of IS and distinguish IS from IE in terms of the focus.
	IS considers networks of businesses as industrial ecosystems.	The purpose of the examination should be mentioned to be meaningful to an IS definition.
		Again, the boundary of an industrial ecosystem - a network of businesses remains unaddressed.

(continued)

Table 2.2 (continued)

Definitions	Key features	Comments
'In our experience, IS is not essentially localised waste and by-product exchanges, nor should it be confused with agglomeration economies or industrial clusters where geographic proximity is a necessary condition' (Lombardi and Laybourn 2012, p. 28).	IS is not just concerned with localised waste and by-product exchanges. Agglomeration economies or industrial clusters are different from IS required networks of collaboration.	This definition is a breakthrough from the definition by Chertow (2000). IS should not be distinguished from IE by the scale of economy, as Chertow (2000) considered that IS was for local and regional economies but not for the global economy.
'…geographic proximity is neither necessary nor sufficient for IS, unlike the concepts of agglomeration economies and industrial clusters, which are explicitly geographically based' (Lombardi and Laybourn 2012, p. 31).	Geographic proximity is not a condition for IS applications.	Adds value by stating geographic proximity is not the condition (restriction) to apply IS. Distinguishes industrial clusters from an IS network of diverse organisations that are traditionally unrelated. Or could IS work on both?
'IS engages diverse organisations in a network to foster eco-innovation and long-term culture change. Creating and sharing through the network yields mutually profitable transactions for novel sourcing of required inputs, value-added destinations for non-product outputs, and improved business and technical processes' (Lombardi and Laybourn 2012, pp. 31–32).	IS fosters eco-innovation and long-term culture change by engaging diverse organisations. IS creates a network for sharing. IS searches novel sourcing of required inputs and value-added destinations for non-product outputs. IS improves business and technical processes.	Specifies novel exchanges. The definition also emphasises mutual economic values for IS engaged companies. However, the role of IS which is to support the achievement of the ultimate goal of IE - developing nearly closed-loop ecosystems, still needs to be explicitly specified.

(continued)

Table 2.2 (continued)

Definitions	Key features	Comments
'IS applies the ecological metaphor of IE to create a collective approach to firms and industries traditionally viewed as separate entities and considers the entire system with regard to the physical exchanges of materials, energy, water and by-products' (Leigh and Li 2015, p. 632).	IS applies the ecological metaphor of IE. IS creates a collective approach. IS considers the entire system.	Emphasises the importance of the IE principle of ecological metaphor required in IS. Emphasises the entire system for IS consideration. The system boundary remains unaddressed.

2.3 Development Histories of Industrial Ecology and Industrial Symbiosis

As we gradually recognised the severity of the long-term negative impact of human industrial activities on the Earth, some of us began to actively compare our industrial systems to biological systems. The learning is to sustain our human activities on the Earth for our generation and future generations. The comparison of our industrial systems and biological systems led to the formal initialisation and development of IE and its related areas, including IS, and their applications.

The development history of IE highlights that IE has gradually developed into a study field (Fig. 2.1), which is an interdisciplinary study field. IE, as a study field, includes interrelated study areas, such as industrial metabolism (IM) and IS (Tibbs 1992). In each of these areas, different approaches and frameworks have been developed to support the achievement the goal of IE, which is developing nearly closed-loop industrial ecosystems. The development history of IE also suggests that IE is part of the Dominant Social Paradigm (DSP), which represents IE's significance in the social science disciplines.

For this book, the history of IE is reviewed from 1989 when Frosch and Gallopoulos (1989) published their famous article entitled 'Manufacturing Strategy' in the Journal of Scientific American. The original title, 'Manufacturing - The Industrial Ecosystem View' proposed by authors, was not accepted. The famous logo of their paper is 'wastes from one industrial process can serve as the raw materials for another' (Frosch and Gallopoulos 1989, p. 144). The paper focused on 'industrial ecosystems' and how industrial systems could mimic biological ecosystems to be sustainable in the long run (Frosch and Gallopoulos 1989). They used three material cycles, the iron cycle, the plastic cycle and the platinum-group-metal cycle as examples to describe how different industries could work together to create nearly closed-loop material exchanges to develop industrial ecosystems. The article did not directly define IE but clearly mentioned that IE was for developing industrial ecosystems and the key feature of an industrial

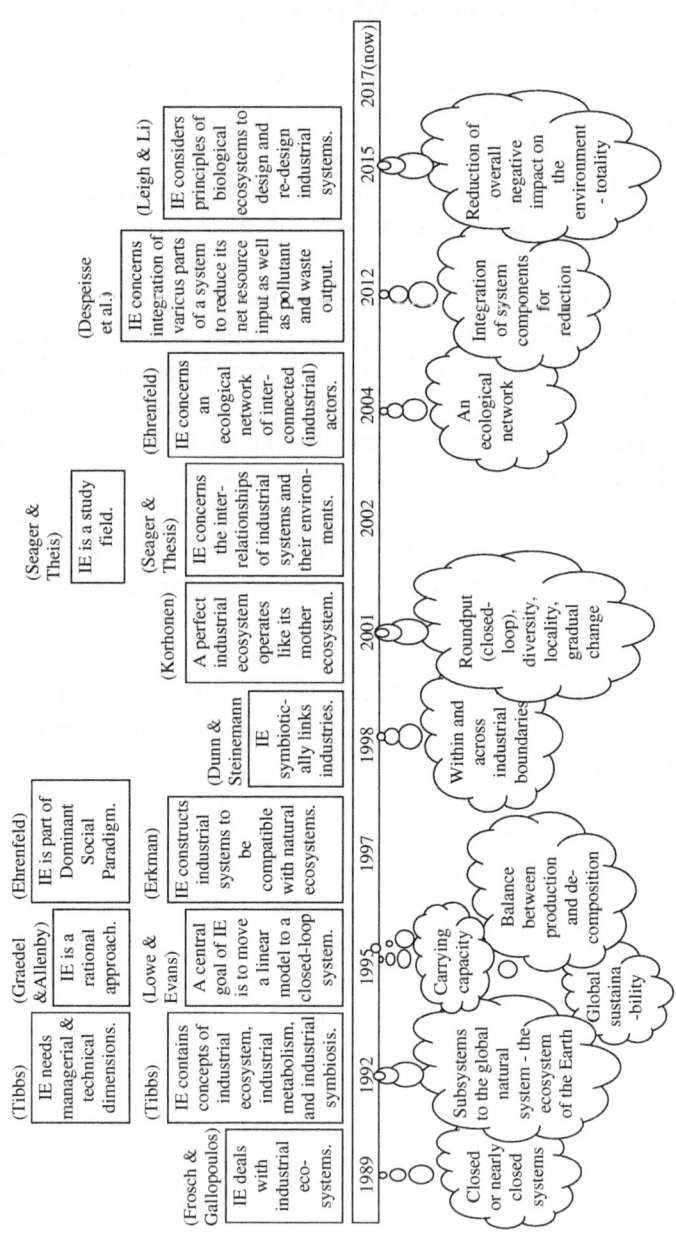

Fig. 2.1 Development history of Industrial ecology

ecosystem was its nearly closed-loop material exchanges. The paper clearly stated that 'manufacturing processes in an industrial ecosystem simply transform circulating stocks of materials from one shape to another; the circulating stock decreases when some material is unavoidably lost, and it increases to meet the needs of a growing population' (Frosch and Gallopoulos 1989, p. 146). The authors emphasised the importance of industries, environmental groups, and individuals working together to contribute to the establishment of industrial ecosystems (Frosch and Gallopoulos 1989). Developing industrial ecosystems is the core of IE, as IE continuously explores different approaches for designing and developing industrial ecosystems across different disciplines.

The development history of IE confirms that the understanding of IE is still yet to be converged. Which features of biological ecosystems an industrial system can mimic to become an industrial ecosystem still needs further exploration. Some common features of industrial ecosystems identified in IE development history are:

- nearly closed-loop material exchanges,
- balance between production and decomposition,
- diversity of industrial units/processes/organisations, and
- totality, which is an extended system view.

An established industrial ecosystem still needs to change continuously in order to adapt to its business environment and its mother ecosystem. The changes should be gradual, allowing the system to regain and maintain the balance over time and space. A set of nearly closed-loop industrial ecosystems are part of their mother ecosystem - the closed-loop ecosystem of the Earth. Locality in biological ecosystems makes the convenience of exchanges between species without the need for travelling through distance. However, when we look into the three material cycles described by Frosch and Gallopoulos (1989), the three material cycles of iron, plastic and platinum-group-metals or their recycle systems can cover a larger geographic area, even globally. This is the same with many other materials for reuse, remanufacturing and recycling, as human development has created the most extensive transportation

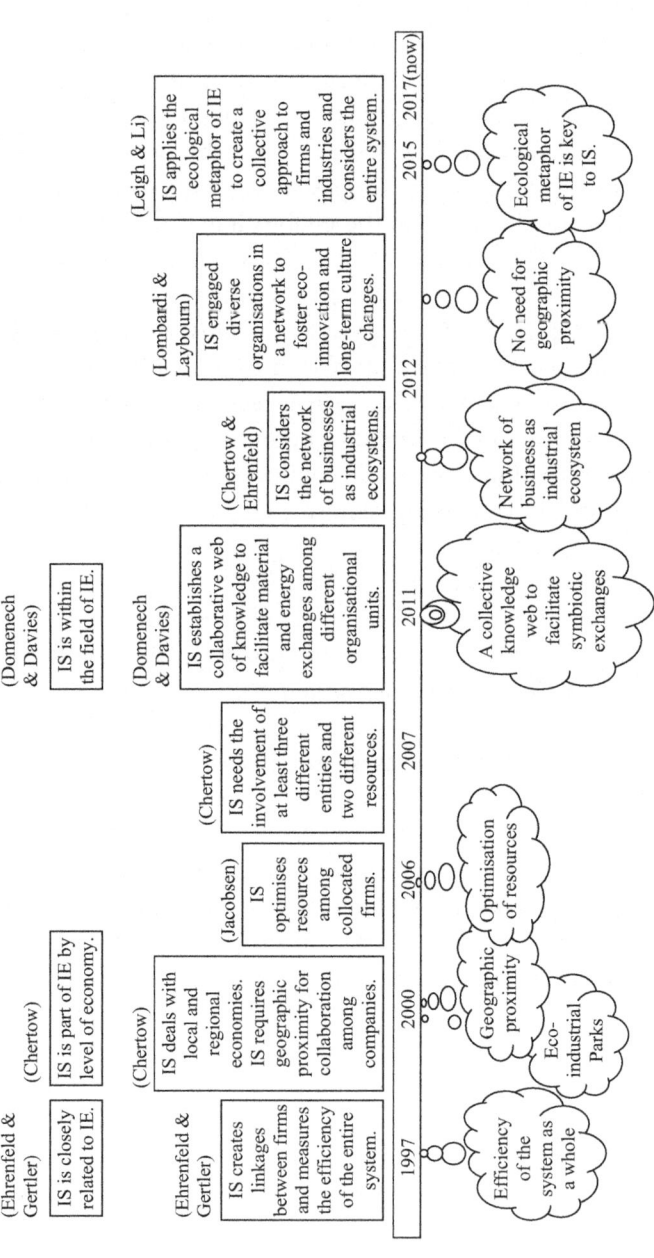

Fig. 2.2 Development history of Industrial symbiosis

system on the planet. Locality should not be a feature for industrial ecosystems to mimic, only if we ignore the globalisation of the industrial world and its advance in transportation technology.

The key question remains the boundary of an industrial ecosystem. The boundary of an industrial ecosystem is up to study purposes, like the boundary of a biological ecosystem. The degree of interaction, which this industrial ecosystem has with other industrial ecosystems and further with the overall ecosystem of the Earth, determines the quantity and rate of material exchanges among them. The quantity and rate of material exchanges within a system influence the level of closed-loop material exchanges of this system and its mother system at an extended system boundary. For example, an industrial ecosystem with a predetermined system boundary might have only 50 per cent of closed-loop material exchange within this system boundary. However, its mother industrial ecosystem, which contains this industrial ecosystem and a few other ones, has higher, say 70 per cent of closed-loop material exchanges. If increasing the level of closed-loop material exchanges within this industrial ecosystem contributes to the increased level of closed-loop material exchanges of its mother industrial ecosystem, there would not be any problem. However, if that is not the case, should we continue increasing the level of closed-loop material exchanges within this ecosystem or should we allow its material exchanges with other sub-industrial ecosystems in order for its mother industrial ecosystem to achieve a higher level of closed-loop material exchanges overall? As the totality is the key concern for IE, the boundary of an industrial ecosystem becomes an issue for the development of industrial ecosystems. The system boundary issue is further reflected in the development history of IS (Fig. 2.2).

The development history of IS (Fig. 2.2) clearly recognises that IS is within the study field of IE (Domenech and Davies 2011). However, there are a number of debatable concepts presented in the IS development history. This includes geographic proximity, self-organisation, and the required number of entities and different resources involved in an IS exchange. Some of these debatable concepts have been brought to attention when reviewing definitions of IS. Hence, no repetition is given here. The emphasis here is that IS focuses on symbiotic relationship establishment for achieving the ultimate goal of IE, which is the development of nearly closed-loop industrial ecosystems. However, the development history of IS does not always reflect this purpose of IE and contains some subjective and misleading concepts.

The term, 'industrial symbiosis' (IS) was mentioned many years prior to 2000 (Tibbs 1992; Lowe and Evans 1995). For example, Tibbs (1992) considered IS as one of the concepts within IE. The classic example of Kalundborg in Denmark has been used to illustrate applications of both IE and IS in the literature (Domenech and Davies 2011; Ehrenfeld and Gertler 1997; Jacobsen 2006). These studies frequently emphasised the unplanned collaborations established in Kalundborg over the years for cross-industrial materials, waste, by-products and energy exchanges in a local community (Domenech and Davies 2011; Ehrenfeld and Gertler 1997; Jacobsen 2006). However, more recent synergies in Kalundborg have been facilitated by the Kalundborg Symbiosis Centre, established in 1996 (Branson 2016; Valentine 2016). In addition, the successful UK national IS programme (NISP) has been a planned and facilitated IS programme, coordinated by NISP centres in different regions and its central NISP team (Mirata 2004; Jensen et al. 2011).

A number of researchers emphasised the importance of diverse industries for symbiotic relationship establishment in IS practice (Chertow 2000; Costa and Ferrão 2010). However, some researchers implied that IS could also be applied within firms (Despeisse et al. 2012; Lehtoranta et al. 2011). Some studies focused on IS alone in terms of its role in establishing symbiotic relationships among different industrial companies without consideration of closed-loop industrial ecosystem development (Chertow and Ehrenfeld 2012; Chertow 2007). Developing symbiotic relationships is an important step towards the closed-loop principle of IE. However, it is the totality of IE which needs to be reflected in developing these symbiotic relationships in relation to IS. The exploration of the relationships between IE and IS and other areas within the study field of IE aids this understanding.

2.4 The Relationship between Industrial Ecology and Industrial Symbiosis and Study Areas within Industrial Ecology

When a study explores or is entitled IS, it often opens with an explanation of IE (Chertow 2000; Domenech and Davies 2011; Lombardi and Laybourn 2012; Van Berkel et al. 2009; Wang et al. 2013). IS cannot or should not exist without consideration of IE. However, IS has its own distinctive focus within the study field of IE and this needs to be emphasised when exploring IS and its applications. Some explanations of IE and IS share great similarity, which consider both IE and IS focusing on flows of materials and energy and applying ecological metaphor (Chertow 2000; Ehrenfeld 1997). This can confuse readers and neglect the important role of IS in achieving the ultimate goal of IE.

Using Chertow's (2000) paper as an example, the definitions for IE and IS are almost identical but distinguished by the scale of economy. In the abstract of the paper, IS was explained thus: 'Industrial symbiosis, part of the emerging field of industrial ecology, demands absolute attention to the flow of materials and energy through local and regional economies' (Chertow 2000, p. 313). The first sentence in the introduction of the same paper states 'The emerging field of industrial ecology [IE] demands resolute attention to the flow of materials and energy through local, regional, and global economies' (Chertow 2000, p. 314). The difference between these two statements is that IE is concerned with local and regional as well as global economies; whereas IS considers local and regional economies but not the global economy. The question is why not? The reason behind this is the misguided use of geographic proximity as an essential condition of IS, and the assumption that physical material exchanges can only take place in local and regional economies, which is certainly untrue. It might be more advantageous in terms of transportation costs and infrastructure for material exchanges locally. Material exchanges in relation to IE and IS should be distinguished from normal business trading material exchanges by their novelty and impact on the reduction of intakes of virgin materials and

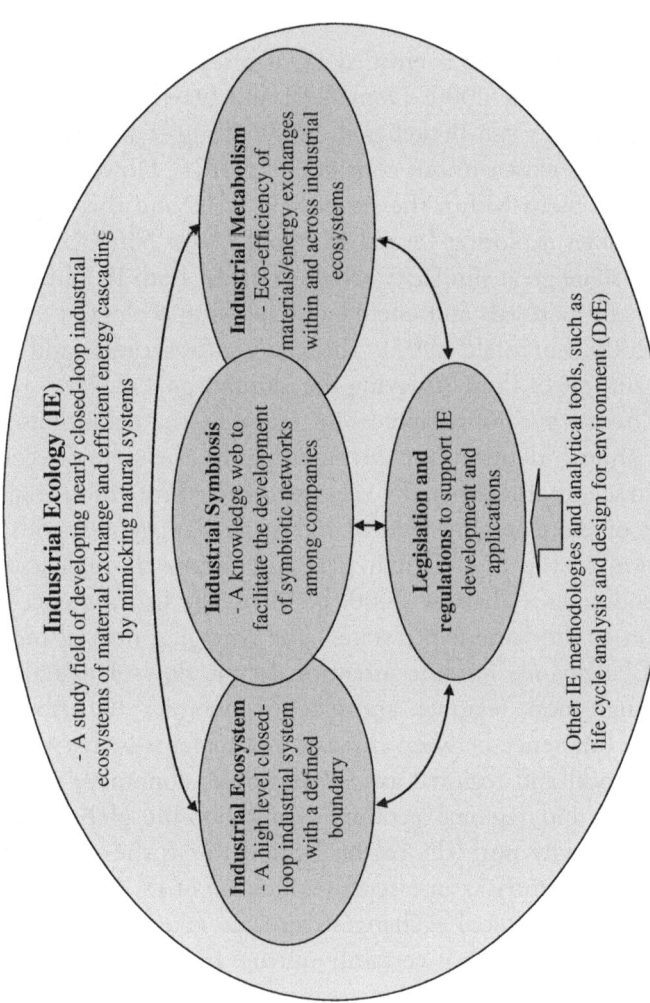

Fig. 2.3 Industrial ecology and its interrelated four study areas

disposals of waste to nature. The real difference between IE and IS lies in their focuses, but not the scale of economy concerned.

Chertow (2000) continued that 'Industrial ecology allows focus at the facility level, at the inter-firm level, and at the regional or global level. Industrial symbiosis occurs at the inter-firm level because it includes exchange options among several organisations' (Chertow 2000, p. 314). But how could IE at the global level not involve any exchange options among several organisations? IS does focus on the inter-firm collaborations for novel materials exchanges to develop industrial ecosystems. However, the inter-firm collaboration can be local, regional, national and international. IE and IS are related by the same principles that IS is part of IE. IS has extended the study field of IE by developing a set of methods to promote the establishment of collaborations between companies among different industries for novel exchanges, regardless of geographic location, to achieve the goal of IE.

The same terms, such as 'ecological metaphor' and 'knowledge sharing and collaboration' have been used to explain both IE and IS through their development and applications in the literature (Chertow 2000; Ehrenfeld 1997). In some cases, IE practices were explained using the term of IS (Ehrenfeld and Gertler 1997). Some IS applications, particularly EIPs, need to consider the level of closed-loop material exchanges which is the core of a symbiotic relationship in a biological world where IE originated (Ehrenfeld and Gertler 1997; Gibbs and Deutz 2005).

In the study field of IE, there are other areas to consider besides IS. One of the areas is industrial metabolism (IM) (Tibbs 1992; Ayres 1989). IM also aims to achieve the ultimate goal of IE, but focuses on the methods that can be deployed to measure and improve eco-efficiency and rates of exchange to allow more profound and effective exchanges. This book does not cover IM in detail (Sect. 3.4 explains some basics of IM). However, we need to highlight the relevance of IM within the study field of IE. In addition, there are studies that have explored legislation and regulations in relation to the development and applications of IE and IS (Malcolm and Clift 2002) (Sect. 3.5 explores this further). This proposes another important area within the study field of IE. Therefore, the current thinking of IE and its study areas are illustrated in Fig. 2.3.

In each of these areas, theoretical concepts and methods for application have been developed, along with successful factors as well as barriers to overcome. The areas are integrated to support the achievement of the IE ultimate goal. For example, legislation and regulations can serve as either successful factors or barriers depending on how the current legislation and regulations have been set and developed in relation to IE applications, including the development of industrial ecosystems, IS and IM. These all can pose challenges to IE and IS applications. The final chapter of this book explores challenges for IE and IS applications in great detail.

IE focuses on the general aspects of industrial ecosystems mimicking suitable features of biological ecosystems and explores the principles of biological ecosystems for industrial systems to be developed into nearly closed-loop industrial ecosystems. IE considers that an industrial ecosystem is part of its mother natural ecosystem. Therefore, the development of an industrial ecosystem needs to support the achievement of the balance of interrelated ecosystems whatever biological ecosystems and industrial ecosystems are concerned. The total balance has to be maintained through gradual but constant changes to adjust each player in the total natural ecosystem of the Earth through every single transformation of materials and energy.

IS emphasises the application of IE principles through consciously establishing symbiotic relationships among industrial entities through developing knowledge webs of novel ideas. Explanations and applications of IS (or claimed as IE) need to consider the principles of IE when highlighting the symbiotic relationship establishment to facilitate the movement of industrial systems towards nearly closed-loop industrial ecosystems of material exchanges and energy cascading.

2.5 Summary

By critically reviewing different definitions of IE and IS, this book defines IE and IS as:

> Industrial Ecology (IE) is an interdisciplinary study field containing interrelated study areas of industrial ecosystem, industrial symbiosis (IS),

industrial metabolism (IM) and legislation and regulations for IE development and applications. IE embraces and develops different approaches, both technical and managerial, to design industrial ecosystems and to transform industrial systems to industrial ecosystems, through mimicking suitable features of biological ecosystems. IE aims to develop nearly closed-loop industrial ecosystems, which are balanced, diverse and gradually changing in feature, in terms of material exchanges and energy cascading.

Industrial Symbiosis (IS) explores ways to establish knowledge webs of novel material, energy and waste exchanges and business core processes to facilitate the development of networks of synergies within and across different companies to support the development of high levels of nearly closed-loop material exchanges and efficiency of energy cascading within and across industrial ecosystems.

The four IE areas should not be applied in isolation as each cannot claim to be IE. It is the integrated effect of applications that leads to the achievement of the ultimate IE goal for developing nearly closed-loop industrial ecosystems. IS explores different ways to establish knowledge webs of novel ideas and a network of potential industrial partners to maximise opportunities of novel exchanges within and among different organisations to fulfil the goal of IE.

IE needs to 'draw in theorists and practitioners from many disciplines or fields that have become separated by the inexorable processes of modernist epistemology' (Ehrenfeld 2004, p. 830). IE is an interdisciplinary study field that proposes challenges for integrating different study disciplines. OM should be supported by IE thinking and considers transformation processes and systems in a closed-loop manner instead of a linear representation. The closed-loop thinking for business decisions can impact significantly on the reduction of the intake of virgin materials from and the disposal of waste to the natural environment.

In the IE and IS development, economic gains as a driven factor for companies to participate in developing symbiotic relationships have been greatly emphasised, such as those described in the classic example of Kalundborg, Denmark. Focusing on economic gains as the primary aim only allows a small scale of randomly occurring IE and IS applications. Environmental gains need to be given much more emphasis in practice. The related areas of IE and their integration in application across disciplines need to be embedded in the development of environmental sustainability. The world needs much greater inputs from IE and IS in order for industrial activities and our living standards to sustain. IE still needs to establish its critical role as part of the DSP to further influence business decisions and culture change (Ehrenfeld 1997). The idea of industrial systems, as a form of industrial ecosystems, being part of the mother ecosystem of the Earth needs to be rooted in our thinking for business planning and decision-making. In order for businesses and industries to prosper in the long run, our business decisions need to follow or mimic more closely the rules of nature. IE and IS contribute to the transformation of this thinking and business practices.

References

Ayres, R. U. (1989). Industrial metabolism. In J. H. Ausubel & H. E. Sladovich (Eds.), *Technology and Environment* (pp. 23–49), Washington, DC: National Academies Press.

Boix, M., Montastruc, L., Azzaro-Pantel, C., & Domenech, S. (2015). Optimization methods applied to the design of eco-industrial parks: A literature review. *Journal of Cleaner Production, 87,* 303–317.

Branson, R. (2016). Re-structuring Kalundborg: The reality of bilateral symbiosis and other insights. *Journal of Cleaner Production, 112,* 4344–4352.

Chertow, M. (2000). Industrial symbiosis: Literature and taxonomy. *Annual Review of Energy and the Environment, 25,* 313–317.

Chertow, M. (2007). "Uncovering" industrial symbiosis. *Journal of Industrial Ecology, 11,* 11–30.

Chertow, M., & Ehrenfeld, J. (2012). Organizing self-organising system, towards a theory of Industrial Symbiosis. *Journal of Industrial Ecology, 16,* 13–27.

Costa, I., & Ferrão, P. (2010). A case study of industrial symbiosis development using a middle-out approach. *Journal of Cleaner Production, 18,* 984–992.

Despeisse, M., Ball, P. D., Evans, S., & Levers, A. (2012). Industrial ecology at factory level—A conceptual model. *Journal of Cleaner Production, 31,* 30–39.

Domenech, T., & Davies, M. (2011). Structure and morphology of industrial symbiosis network: The case of Kalundborg. *Procedia Social and Behavioral Sciences, 10,* 79–89.

Dunn, B. C., & Steinemann, A. (1998). Industrial ecology for sustainable communities. *Journal of Environmental Planning and Management, 41,* 661–672.

Ehrenfeld, J. (1997). Industrial ecology: A framework for product and process design. *Journal of Cleaner Production, 5,* 87–95.

Ehrenfeld, J. (2004). Industrial ecology: A new field or only a metaphor? *Journal of Cleaner Production, 12,* 825–831.

Ehrenfeld, J., & Gertler, N. (1997). Industrial ecology in practice: The evolution of interdependence at Kalundborg. *Journal of Industrial Ecology, 1,* 67–79.

Erkman, S. (1997). Industrial ecology: An historical view. *Journal of Cleaner Production, 5,* 1–10.

Frosch, R. A., & Gallopoulos, N. E. (1989). Strategies for manufacturing. *Scientific American, 261*(September), 144–152.

Gibbs, D., & Deutz, P. (2005). Implementing industrial ecology? Planning for eco-industrial parks in the USA. *Geoforum, 36,* 452–464.

Graedel, T. E., & Allenby, B. R. (2003). *Industrial Ecology* (2nd ed.). Englewood Cliffs: AT&T and Prentice Hall (1st ed. 1995).

Heeres, R. R., Vermulen, W. J. V., & de Walle, F. B. (2004). Eco-industrial park initiatives in the USA and the Netherlands: First lessons. *Journal of Cleaner Production, 12,* 985–995.

Jacobsen, N. B. (2006). Industrial symbiosis in Kalundborg, Denmark: A quantitative assessment of economic and environmental aspects. *Journal of Industrial Ecology, 10,* 239–255.

Jelinski, L. W., Graedel, R. A., Laudise, R. A., McCall, D. W., & Patel, C. K. N. (1992). Industrial ecology: Concepts and approaches. *Proceedings of the National Academy of Sciences of the United States of America, 89,* 793–797.

Jensen, P. D., Basson, L., Hellawell, E., Bailey, M. R., & Leach, M. (2011). Quantifying 'geographic proximity': Experiences from United Kingdom's National Industrial Symbiosis Programme. *Resources, Conservation and Recycling, 55,* 703–712.

Korhonen, J. (2001). Four ecosystem principles for an industrial ecosystem. *Journal of Cleaner Production, 9,* 253–259.

Lehtoranta, S., Nissinen, A., Mattila, T., & Melanen, M. (2011). Industrial symbiosis and the policy instruments of sustainable consumption and production. *Journal of Cleaner Production, 19,* 1865–1875.

Leigh, M., & Li, X. (2015). Industrial ecology, industrial symbiosis and supply chain environmental sustainability: A case study of a large UK distributor. *Journal of Cleaner Production, 106,* 632–643.

Lombardi, D. R., & Laybourn, P. (2012). Redefining industrial symbiosis. *Journal of Industrial Ecology, 16,* 28–37.

Lowe, E., & Evans, L. (1995). Industrial ecology and industrial ecosystems. *Journal of Cleaner Production, 3,* 47–53.

Malcolm, R., & Clift, R. (2002). Barriers to Industrial Ecology: The Strange Case of "The Tombesi Bypass". *Journal of Industrial Ecology, 6* (1), 4–7.

Mirata, M. (2004). Experiences from early stages of a national industrial symbiosis programme in the UK: Determinants and coordination challenges. *Journal of Cleaner Production, 12,* 967–983.

Park, J. M., Park, J. Y., & Park, H.-S. (2016). A review of the eco-industrial park development program in Korea: Progress and achievement in the first phase, 2005–2010. *Journal of Cleaner Production, 114,* 33–44.

Seager, T. P., & Theis, T. L. (2002). A uniform definition and quantitative basis for industrial ecology. *Journal of Cleaner Production, 10,* 225–235.

Tian, J., Liu, W., Lai, B., Li, X., & Chen, L. (2014). Study of the performance of eco-industrial park development in China. *Journal of Cleaner Production, 64,* 486–494.

Tibbs, H. (1992). Industrial ecology, an environmental agenda for industry. *Whole Earth Review,* winter, 4–19.

Valentine, S. V. (2016). Kalundborg Symbiosis: Fostering progressive innovation in environmental networks. *Journal of Cleaner Production, 118,* 65–77.

Van Berkel, R., Fujita, T., Hashimoto, S., & Fujii, M. (2009). Quantitative assessment of urban and industrial symbiosis in Kawasaki. *Japan, Environmental Science, 43,* 1271–1281.

Wang, G., Feng, X., & Chu, K. H. (2013). A novel approach for stability analysis of industrial symbiosis systems. *Journal of Cleaner Production, 39,* 9–16.

3

Industrial Ecology Applications in the Four Areas

Abstract Industrial Ecology (IE) can be applied in four interrelated study areas: industrial ecosystem, Industrial Symbiosis (IS), industrial metabolism (IM) and legislation and regulations for IE applications. Different methods can be used to determine the boundary of an industrial ecosystem: material-based, product-based and geographic-based. IS applications have shifted from a self-organising to planned or facilitated practice. IM establishes its position within IE in quantifying the efficiency and rates of material, waste and energy exchanges over the total corresponding flow to evaluate the closed-loop status of an industrial ecosystem. Legislation and regulations on waste need to reflect the IE's view of waste as resources. The integration of these four areas is critical to the success of IE applications to fulfil its potential to improve environmental sustainability.

Keywords Applications of Industrial Ecology · Industrial ecosystem boundaries · Self-organised versus planned or facilitated Industrial Symbiosis · Efficiency and rates of resource flows in industrial metabolism · Legislation and regulations for Industrial Ecology applications

3.1 Introduction

The debate in the Industrial Ecology (IE) literature has focused on whether IE has been applied in practice, or remains as a metaphor (Ehrenfeld 2004). The exploration of IE primarily remains theoretical in general and focuses on areas for further development and possible applications. IE contains four study areas: industrial ecosystem, Industrial Symbiosis (IS), industrial metabolism (IM) and legislation and regulations for IE applications. In each of these four study areas, concepts, frameworks, tools and approaches have been developed and integrated to guide the practice. The integration of these four areas is critical to the success of IE applications in achieving its ultimate goal - high levels of closed-loop material exchanges and energy cascading of industrial ecosystems.

Applied IE requires programmes to integrate managerial and technical elements (Tibbs 1992). The managerial side of IE provides businesses/organisations with the direction for strategic planning and decision-making and also for deploying different analytic tools to guide the IE practice. The management side of IE considers:

- Determining the system boundary and mapping industrial (eco) systems,
- Identifying potential innovative flows of material exchanges and energy cascading within and across companies and industries, or industrial ecosystems,
- Developing symbiotic relationships among companies and industries,
- Improving IM - the efficiency and rate of resource flows of innovative material exchanges and energy cascading, and
- Identifying required legislation and regulations on all scales, including consumers, companies, industries, supply chains and beyond, to support IE applications.

The elements on the managerial side are particularly relevant to operations management (OM) and also to other management disciplines, not just for environmental management. OM is concerned with the process of delivering goods and services efficiently and effectively, from operations strategy formulation, product and process design and development, technology deployment, production planning and control, and continuous improvement of all of these areas. OM focuses on the analysis of business and operations capacity, resource and capability and proposes actions for improvement. The principle of nearly closed-loop material exchange and high efficiency of energy cascading in IE needs to be integrated with these OM elements to have a profound practical impact on environmental sustainability, and further impact on economic and social sustainability. The managerial side of IE needs to be integrated with technical elements of IE to ensure successful applications. The technical side of IE includes:

- Specific eco-features required for green products and green processes,
- Methods for eco-design (such as detailed technical elements and solutions in design for environment (DfE)),
- Green technology development and deployment,
- Methods for dematerialisation, and
- Scientific methods for data gathering and analysis of the status of industrial (eco)systems.

In order to implement these technical elements successfully, effective management is critical. For example, management can focus on the improvement in efficiency and effectiveness of product and process design as well as technology development and deployment. The key is to integrate managerial and technical elements, as the pursuit of these elements in isolation is insufficient to have a profound impact on environmental sustainability.

This chapter explores the following four aspects in relation to IE applications, focusing on managerial aspects:

1. Determining the boundary of an industrial ecosystem,
2. Developing symbiotic relationships (IS),
3. Improving industrial metabolism (IM), and
4. Aligning legislation and regulations with a long-term perspective of industrial (eco)system evolution.

Over time and space, achieving advances in the above four aspects helps develop nearly closed-loop industrial ecosystems, which are integrated systems within the natural ecosystem of the Earth with reduced impact on the natural environment. This can dramatically reduce the need for extracting virgin materials from natural systems and disposing waste and emissions to our planet and atmosphere.

3.2 Determining the Boundary of an Industrial Ecosystem

An industrial ecosystem is an industrial system of nearly closed-loop material exchanges and high efficient energy cascading. An industrial ecosystem has a predefined boundary for study purposes. Determining an industrial ecosystem boundary is a complex task and requires extended system thinking, as each industrial (eco)system does not stand alone without interacting with other systems within the ecosystem of the Earth. There are three interrelated approaches to determine the boundary of an industrial ecosystem in the literature: material-based, product-based and geographic-based (Frosch and Gallopoulos 1989; Korhonen 2002). For the material-based approach, the flow of an individual material, such as iron, steel, polyvinyl chloride, and platinum group metals, can be considered an industrial ecosystem for exploration (Frosch and Gallopoulos 1989). For a product-based approach, an industrial ecosystem follows activities along a product life cycle, from material acquisition to product after-use, for example, the product of paper for printing (Korhonen 2002), for obtaining wood pulps to deinking and recycling of used printing paper. For a geographic-based approach, an industrial ecosystem can be just a physical boundary of a factory facility, an (eco)industrial park, a community or region, such as the regional community of Kalundborg in Denmark (Côte and Hall 1995; Korhonen 2002). Each of these approaches has its own features. The key features of these approaches for determining the boundary of an industrial ecosystem are summarised in Table 3.1.

Each of these three approaches for determining the boundary of an industrial ecosystem has its own advantages and disadvantages in terms

3 Industrial Ecology Applications in the Four Areas

Table 3.1 Key features of approaches for determining industrial ecosystem boundary

Approaches	Key features
Material-based	• Focusing on the flow of a single material, from its extraction, usage, recycling and transformation in different products over time and space • Collaboration among industries, recycling companies and in some cases consumers • Single dimension of a material to focus on and multiple participations of companies and industries
Product-based	• Focusing on different material flows of a single product along the product's life cycle and after-use, including material acquisition, production, distribution, consumption and after-use recycling, remanufacturing or reuse • Collaborations among industries, consumers and recycling companies • Multiple dimensions of materials used to make a product, multiple participations of companies along the product supply chain and recycling companies, and multiple dimensions of consumer involvement
Geographic-based	• Focusing on the recycling of industrial waste and by-products (industrial residues) • Collaboration among different industries for novel material exchanges and energy cascading, and recycling companies • Multiple dimensions of materials, products and companies but focusing on a geographically defined area

of ease of use and impact. Table 3.2 summaries some advantages and disadvantages of these three approaches. In practice, these approaches can be combined in use under different circumstances, with typically one approach being dominant.

The material-based approach has been used over centuries even though the association of this approach with industrial ecosystems has only been made in recent decades. The 'recycling' or 'refinery' of precious metals, such as gold and platinum group metals, is one typical example. As gold and platinum group metals carry considerable economic value, great effort has been made to recover them from one product in order to be used or made into a different product. The recycling of iron and steel also has a long history. There is almost no boundary at all as the transformation continues and some quantity might be lost in use and

Table 3.2 Advantages and disadvantages of approaches for determining the industrial ecosystem boundary

Approaches	Advantages	Disadvantages
Material-based	• A high level of nearly closed-loop material exchanges is achievable through material extraction, production, usage and recycling/transformation. • Focus on recycling and transformation of a material.	• Lack of ownership of the material along its flow through different transformations. • The entire closed-loop of the flow of a material is impossible to capture, as the system boundary never ends.
Product-based	• A high level of closed-loop exchanges of main materials of a product is highly achievable through the product life cycle including product after-use. • Focus on recycling, reuse and remanufacturing possibilities of a product and its components.	• Challenging to define the boundary, as the boundary is not visible geographically and a product can contain a significant number of materials to be included. • Complexity in development due to the involvement of individual consumers.
Geographic-based	• Easy to define the boundary. • Focus on collaborations between different companies and industries for novel industrial waste/by-products exchanges and energy cascading.	• End-life product recyclability is not considered. • Levels of closed-loop material and product flows are low, as the recycling of after-use products is often not considered.

transformation processes; whilst new source materials are added to their circulations through extracting more from nature. For other materials, such as natural gypsum, which is not expensive to extract, less effort has been put into converting sulphur from coal-burning power plants into synthetic gypsum to be used for construction (Ayres 1994). The material-based approach focuses on the flow of a single material. The approach gives limited consideration to industrial waste and by-products associated with the transformation processes. The material-based approach can be used to consider a few interrelated materials as a material-web of different material flows. This increases complexity in mapping and analysing the industrial ecosystem which may contain a large number of material flows.

The product-based approach considers the entire product life cycle including reuse, remanufacturing and recycling of the materials in a product after its use. This can lead to high levels of nearly closed-loop material flows of a product, if product reuse, remanufacturing and recycling are successful, with the integration of DfE. The product-based approach focuses on a single product life cycle including its after-use. Like the material-based approach, the product-based approach neglects possibilities of collaborations among different industries for novel exchanges of industrial waste and by-products. Complexity of the product-based approach is due to the essential involvement of consumers and sheer number of materials, in some cases, involved in a product. However, this can be reduced by restricting the number of materials in a product to be considered in an analysis along the product life cycle and its direct after-use. Consumer involvement leads to the need for recycling companies to act as the waste collector and processor. Under current business legislation and regulations, companies producing products have little responsibility and control over their products' after-use. This needs to be changed but requires new legislation and regulations to consider both companies and consumers' benefits and also their responsibilities for products produced and used. The approaches such as extended producer responsibility (EPR) (Lifset et al. 2013; Gui et al. 2016; Xiang and Ming 2011) and product stewardship (Lewis 2005; Rogers et al. 2010) support this development.

The geographic-based approach typically sets the boundary of industrial ecosystems within an industrial park, a local community, or a region, even though this approach can also be used for a factory facility. The boundary is easily determined as it is geographically based. The analysis of this type of industrial ecosystem focuses on opportunities for recycling industrial waste, waste water and energy, and by-products through collaborations. Collaborations for novel exchanges of recyclable waste, including waste water and energy among different industries, are given priority to increase the level of waste and by-products as resources re-entering into a different process within a geographically defined boundary. It is just like waste from one process becoming feedstock to another (Frosch and Gallopoulos 1989), but focusing on developing collaborations within a geographically defined boundary. This increases the quantity of material and energy flows within this geographically defined area. Therefore, it reduces the need for resource inputs from, and pollution and waste emissions to the external environment, and/or increases business opportunities within the geographic area.

In some cases, this type of system boundary has to be extended for analytical purposes to include inflow and outflow to the geographic area to have valuable conclusions (Lu et al. 2015). One disadvantage of this geographic-based industrial ecosystem is that it does not usually consider inflow, such as material and energy suppliers from the outside of this industrial ecosystem, and outflow, such as products and waste emissions to outside of this area. Examples of this type of ecosystems include the industrial ecosystem at Kalundborg, Denmark and many eco-industrial parks (EIPs) around the world. Using the industrial ecosystem of Kalundborg as an example, the external coal supplier to the power station and the destination and after-use recycling of the plaster-boards are not considered. The key contribution of using a geographic-based industrial ecosystem for analysis is to identify opportunities for collaboration among different companies and industries to exchange industrial production waste and by-products, waste waters and energy to reduce waste and emissions by direct industrial activities.

These three approaches for determining the industrial ecosystem boundary are interrelated and can be used simultaneously. However, combining two approaches can increase the complexity in mapping an industrial ecosystem, and can also raise conflicts in decision-making in some circumstances (Korhonen 2002). It is important to start from an industrial ecosystem on a small scale to form its establishment before extending the industrial ecosystem boundary. The analysis of an industrial ecosystem with a boundary and further with an extended boundary supports the way forward to create an increasingly profound impact on the reduced intake of virgin materials and the reduced waste and emissions to the natural environment by industries over time and space.

The boundary of an industrial ecosystem is not fixed. It can and should be extended when new novel material exchange relationships with other companies have been developed and considered in the industrial ecosystem analysis. The use of a single approach either focusing on a single material, product or small geographic area is a good starting point. This can be followed by extending this system boundary when the initial industrial ecosystem has been firmly established in the analysis and practice and the extension of it becomes feasible and beneficial to all participating companies including new participating companies. The practical elements of establishing symbiotic relationships within and across companies and different industries are explored in the next section.

3.3 Developing Symbiotic Relationships

Continuous exploration to identify and develop symbiotic relationships within and across companies and also within and across industries, for novel material exchanges and energy cascading, is an important area for IE applications.

IS explores ways to establish knowledge webs of novel material, energy and waste exchanges and business core processes to facilitate the development of networks of synergies within and across different companies to support the development of high levels of nearly closed-loop material exchanges and efficiency of energy cascading within and across industrial ecosystems. IS collaborations differ from other business collaborations by creating **novel/innovative** material exchanges and energy cascading between business processes, companies and industries. The novel exchanges can be from one process to another, within a single company or industry, and across different companies or industries, but all ultimately creating higher levels of closed-loop material exchanges and efficiency of energy cascading within and across industrial ecosystems. The establishment of knowledge webs to facilitate novel physical material exchanges and high efficient energy cascading is critical to IS practice and development (Lombardi and Laybourn 2012).

Initially, the IS literature emphasised the importance of spontaneity in the development of symbiotic relationships among companies. Economic-driven factors, instead of environmental focused drivers for the establishment and development of these symbiotic relationships, have been heavily emphasised (Tibbs 1992; Gibb and Deutz 2005). The success of deliberation and consciousness of developing IS relationships was debated as well as the transferability of spontaneous symbiosis relationship establishment (Gibb and Deutz 2005). Some proposed that government investment should be given to those companies that have set out to establish IS exchanges, rather than investing in the search for new partners (Heeres et al. 2004). The question is that, if the support and investment only focus on fostering existing companies in their development, how could an IS network be further developed? Without continuously involving more companies and industries in the IS network, the development of IS networks for achieving the ultimate goal of IE remains relatively limited, compared to the entire scale of the industrial economy.

The linear transformation representation of industrial processes, which is currently used in OM, will continue to dominate the industrial world and the achievement of environmental sustainability remains a challenge. 'Self-organising' might be one way to foster an IS network to evolve, such as the Kalundborg IS regional community during its early years (Chertow 2000). However, this type of IS practice has very limited transferability. 'Planned development' or 'deliberate attempts' (Jensen et al. 2011), such as the development of Kalundborg IS community in recent years, the UK national IS programme (NISP), and eco-industrial parks (EIPs) around the world, can create positive progress of IS development in local communities, industrial parks, regions, nations and beyond (Jensen et al. 2011; Zhang et al. 2015).

Establishing new symbiotic exchanges within an established industrial ecosystem and adding new companies to it benefit the development of the IS network and exchanges. Adding new partners to expand the current IS network to facilitate its further development and continuously transferring industrial companies and industrial parks to industrial ecosystems are ways to apply IS (Velenturf 2016). Besides increasing the number of IS partners in an IS network, increasing quantity and efficiency of the current exchanges can also contribute to the development of higher levels of closed-loop material exchanges and efficiency of energy cascading within and across industrial ecosystems to some extent. IM, as one area in IE, quantifies flows of symbiotic exchanges in IS applications. It is the relativity of improvement in the IS network, in terms of the number of partners and the quantity of exchanges over time and space, that matters.

There are a number of **ways** to develop symbiotic relationships:

- Develop knowledge webs to facilitate different types of symbiotic material exchanges and energy cascading (Lombardi and Laybourn 2012),
- Establish the IS centre to organise regular workshops to promote IS thinking and identify potential partners for new symbiotic relationships (Mirata 2004 - the NIPS coordinating centre; Shi et al. 2010 - The TEDA Eco-Centre in Tianjin, China),
- Encourage research and collaborations among education institutions, companies and industries, IS centres, consumers and recycling companies (Heeres et al. 2004; Park et al. 2016),
- Develop information technology to support knowledge exchanges and database development to facilitate IS relationship establishment (Jensen et al. 2011),
- Develop governmental policies, legislation and regulations as well as investment to promote companies for initial involvement in symbiotic exchanges (Park et al. 2016 - the first phase of the EIP transformation programme in Korea), and
- Promote regional and national IS programmes (Mirata 2004; Jensen et al. 2011 - The UK NISP).

As a planned or facilitated IS development, which is more transferrable, the NISP in the UK offers a number of specific ways in practice. The two distinctive ways in the NISP are: by-passing the trust-building stage among companies through using a trusted programme team, and developing a large database of companies for building synergies (Jensen et al. 2011; Mirata 2004). The literature on EIP development also has some implications for facilitating the establishment of IS relationships. The involvement of participating companies at the early stage of EIP development and informal interaction and communication channels for trust building are essential for these planned IS developments to be successful over time.

Nevertheless, specific ways for facilitating and fostering the establishment and development of IS relationships are still limited and fragmented in the literature. Learning from other disciplines, in terms of the establishment and development of relationships among companies by overcoming barriers, can be beneficial to IS development and applications. Hence, further research in this area is required.

Chapter 4 explores the development of IS relationships and networks in more detail through critically reviewing different types of IS applications in the literature.

3.4 Improving Industrial Metabolism

IM is an important concept and study area within IE. IM has been applied in practice, for example, to explore and quantify carbon metabolism for eco-industrial parks (Lu et al. 2015) and to assess IM of sulphur in a Chinese fine chemical industrial park (Tian et al. 2012). However, the concept has seldom been discussed and definitions are inconsistent (Wassenaar 2015). In addition, an exploration of ways to improve IM is very limited. This section begins with clarification of the concept of IM, but focuses on applied elements for improving IM to achieve the ultimate goal of IE from a managerial perspective. This section does not intend to go into great detail on IM, but to introduce IM as one area within the field of IE.

The term 'metabolism' in 'industrial metabolism' originated from biology and ecology (Ayres 1994; Lu et al. 2015), or might be simultaneously proposed in both biology/ecology and sociology (Fischer-Kowalski 2003). The debate here is not where and when 'metabolism' in the term 'industrial metabolism' originated. However, it is certain that industrial systems can learn from biology and ecology in a number of aspects to be more environmental friendly. This also applies to IM within the field of IE.

In biology and ecology in the nineteenth century, 'metabolism was defined as an exchange of energy and substances between organisms and the environment' by Moleschott (Fischer-Kowalski 2003, p. 38). This definition of metabolism in biology and ecology is very similar to the overall concern of IE today and the difference is that IE considers industrial ecosystems instead of organisms. The definition was useful at that time as ecology and the concept of metabolism were evolving. When ecology has evolved and IE has developed and learned from the principles of ecology, IM and other separate, but integrated areas, have gradually formed to focus on specific aspects to develop industrial ecosystems within IE. The definition of metabolism by Moleschott (Fischer-Kowalski 2003) is no longer specific enough to describe IM.

Ayres (1994, p. 23) stated that 'the metabolism of industry is the whole integrated collection of physical processes that convert raw materials and energy, plus labour, into finished products and wastes in a (more or less) steady-state condition'. This definition contributes to the understanding of IE development, though again it is relevant to IE overall, but not specifically to IM. This key aspect associated with IM is the eco-efficiency of conversion and transformation of a process or a system in relation to input resources, product outputs and recyclable waste and non-recyclable waste. Recyclable waste and non-recyclable waste, as well as input from recycled materials or virgin material, need to be distinguished to define, describe and measure IM.

More recently, Wassenaar (2015) proposed a definition for IM as 'human mediated matter change for sustaining a productive system's economic activity' (p. 722) and also considered IM as 'a subset of a complex system of interconnected transformative processes across all scales of life: the metabolic network' (p. 715). The definition by Wassenaar (2015) was very abstract and general and, therefore, very limited for practical use. In addition, this definition might explain what IM should include in general, but does not explain with which IM is exactly concerned.

Simple ratios to measure IM and factors impacting on the increase of this rate can support its applications and effectiveness of an application to support the achievement of the ultimate goal of IE.

Eco-efficiency of an exchange along a flow, a process, or for an entire industrial ecosystem needs to be specified in the definition of IM. IM is concerned with flows (pathways), exchanges and transformation of recycled and non-recycled materials and energy, and recyclable and non-recyclable products and waste. IM particularly focuses on quantifying and improving eco-efficiencies of these exchanges and transformations between resources, recyclable and reusable products, and recyclable waste along flows (pathways) or processes within an industrial ecosystem. The study of IM aims to work towards zero non-recycled waste and emissions of an industrial ecosystem.

The relevance of resource use efficiency in relation to IM was emphasised by Ayes (1989). This resource use efficiency in relation to IM includes recyclable waste and by-products use efficiency. These efficiencies in relation to IM evaluate the system status and identify weak links along the flows (pathways). The improvement in IM results in a reduced need for virgin materials, reduced loss of material exchanges, and increased recyclability of industrial waste, by-products, and end-life products.

Either for part of an industrial ecosystem, the entire industrial ecosystem, or an extended industrial ecosystem, efficiencies in relation to IM within IE can be measured by industrial metabolic rates. The industrial metabolic rate can be used to evaluate the status of geographic-based industrial ecosystems as well as material- or product-based industrial ecosystems.

The following factors need to be specified and **quantified** and relevant data need to be collected to facilitate the evaluation and quantification of IM using industrial metabolic rates for the industrial ecosystem concerned:

- Total input resources from the external to the industrial ecosystem,
- Recycled input resources and virgin resources from the external to the industrial ecosystem,
- Recycled input resources from internal circulation within the industrial ecosystem

- Non-recyclable industrial waste and by-products,
- Recyclable industrial waste and by-products,
- Total product output,
- Recyclable product (after-use) output to consumers (either within or outside of the industrial ecosystem), and
- Non-recyclable products (after-use) output to consumers (either within or outside of the industrial ecosystem).

Certain ratios generated from various meaningful combinations of inputs and outputs from the above factors can be used to quantify IM in an industrial ecosystem, which evaluates the status of industrial ecosystems in terms of the closed-loop level. The ratios can be used to make comparisons between different industrial ecosystems or the same industrial ecosystem over time. This gives opportunities to identify new pathways or flows which have not yet been established but are important (Lu et al. 2015). By establishing certain missing direct material and energy flows in the system, the overall metabolic rate of the system can be improved to a higher level of closed-loop material exchanges and efficiency of energy cascading in this industrial ecosystem. This requires the integration of IM and IS.

An industrial process or ecosystem with a high industrial metabolic rate generates less industrial non-recyclable waste and highly utilises recycled input resources, both from internal and external sources. A highly efficient closed-loop material exchanges and energy cascading industrial infrastructure has a high industrial metabolic rate and has the capability to take recycled inputs and generate almost no non-recyclable waste, including both industrial waste and product after-use waste.

3.5 Aligning Legislation and Regulations to Support Industrial Ecology Applications

Companies have experienced difficulties to reuse or recycle their industrial waste or bring back unused or unwanted materials/products from their customers to reuse, remanufacture or recycle, due to restriction in the existing legislation and regulations on waste (Leigh and Li 2015;

Park et al. 2016). The legislation does not always support 'waste' as the feedstock for another process and the procedure of handling industrial waste requires a complex process of documentation. If the procedure has been neglected, serious cases can lead to a criminal offence for company directors and managers (Malcolm and Clift 2002). As the legal response time is long and IE is a novel study field, the existing legislation and regulations do not reflect the understanding of 'waste' in IE and do not always make reuse or recycling of industrial waste possible or straightforward. Some interpretations of 'waste' in business court cases have been inconsistent and illogical, which presents a legal barrier for IE applications in practice (Malcolm and Clift 2002).

In order for IE to prosper and support the development of environmental sustainability with its full strength, relevant legislation and regulations, nationwide or internationally, need to be in line with IE principles towards waste and waste management. IE does not view 'waste' as 'waste', but a type of resource along a 'closed-loop' product life cycle, or the full product life cycle, including product after-use stage. IE works towards zero-disposals. However, the current legislation in the European Framework Directive defines waste as 'any substance or object which the holder discards or intends to discard' (cited in Malcolm and Clift 2002, p. 4), leading to different interpretations of 'waste' in industrial court cases, including whether a recovery process is required, or whether it poses a burden to the company (Malcolm and Clift 2002). If it is considered as 'waste', the legislation on waste is applicable and in some cases, the 'waste' cannot be re-entered to a different process and, therefore, a symbiotic relationship cannot be established. This creates difficulty for the reuse or recycling of industrial waste in a number of cases, leading to industrial court cases and inconsistent decisions (Malcolm and Clift 2002). 'Perhaps legislative and policy analysis should be a "required core" subject in industrial ecology. But, until such an integrated approach is refined and adopted in the legal system, then industry will have to shoulder the burden of the waste enforcement regime—another barrier to the application of industrial ecology principles' (Malcolm and Clift 2002, p. 6).

This book has brought the attention on the importance of legislation and regulations on waste management in the study field of IE. However, a detailed exploration of legislation and regulations in relation to waste management is beyond the scope of this book. Rather, this book suggests that IE researchers need to explore ways to increase awareness and understanding of IE and work with policy makers to align legislation and regulations on waste with the principle of IE to support IE applications and future development, to fulfil its fundamental role in improving environmental sustainability.

3.6 Summary

This chapter has focused on the applied elements of IE through exploring applications in four areas. Table 3.3 summaries key applied aspects of each of the four areas.

Three approaches for determining industrial ecosystem boundaries are presented in the area of industrial ecosystem. The three approaches are material-based, product-based and geographic-based. Each approach for determining industrial ecosystem boundaries has advantages and disadvantages, in terms of industrial ecosystem development and ease of use. The material-based approach focuses on the flow of a material and its transformation along different products. The product-based approach considers flows of materials associated with a product along its entire life cycle including product after-use. For the material-based and product-based approaches, it can be challenging to define the boundary of an industrial ecosystem. The geographic-based approach determines the boundary of an industrial ecosystem based on a defined geographic area and focuses on industrial waste, by-products and energy. It is relatively straightforward to use the geographic-based approach to determine the boundary of an industrial ecosystem, but less likely to achieve a high level of closed-loop material exchanges within it. Determining the boundary of an industrial ecosystem is for study purposes and, therefore, the boundary of an industrial ecosystem should be flexible to change. These approaches need to be used in conjunction in order to support the achievement of the ultimate goal of IE. In addition, mapping flows (pathways) of materials, waste and energy as well as changes over time and space and levels of

Table 3.3 Key aspects for applications in each of the four IE areas

Areas in IE	Industrial ecosystem	Industrial symbiosis	Industrial metabolism	IE legislation and regulations
Key aspects	• Boundary and extended boundary for study purposes • Flows/pathways of materials, water and energy • Level of the nearly closed-loop material exchanges • Level of efficiency of energy cascading	• Knowledge webs for novel material, waste and energy exchanges • The role of IE centres • Symbiotic networking • Information technology for databases to support IS development • Culture changes	• Efficiency of materials, waste and energy exchanges among sub-sets within an industrial ecosystem • Efficiency of exchanges of the industrial ecosystem with its environment • Quantification of the level of closed-loop internal and external flows	• The legal system response time • Consistency and logic in definition and interpretation of 'waste' • Alignment of legislation and regulations on waste with the principle of IE

Integration of these aspects across the four IE study areas

efficiency of these exchanges, are also considered in the area of industrial ecosystem.

IS focuses on symbiotic relationship establishment, through developing knowledge webs and establishing coordinating centres to foster 'facilitated', 'planned' or 'deliberate' IS development. The self-organising IS was in evidence at the early stage of IS development, but it has limited transferability, and in most IS programmes, self-organising is no longer the case. We cannot simply 'wait' to let our current industrial world transform itself into an eco-industrial world. Facilitated or planned IS development, with the assistance of an IS centre, is widely adopted in today's IS programmes, including current development of the Kalundborg IS community. Adopting a planned characteristic in IS applications improves the transferability of IS applications, as well as accelerating the progress of transforming the current industrial world to an eco-industrial one.

IM measures eco-efficiencies of symbiotic exchanges and identifies ways to improve the IM of an industrial ecosystem. IM quantifies eco-efficiencies of the flows (pathways) of materials, waste and energy identified within an industrial ecosystem and across different industrial ecosystems. An industrial ecosystem with a high IM generates less total waste, particularly less non-recyclable waste and a high level of closed-loop material exchanges and energy cascading within this industrial ecosystem.

Legislation and regulations on waste need to take the principle of IE into consideration and need not view waste as waste, but as resources along a 'closed-loop' product life cycle. This 'closed-loop' product life cycle focuses on the development of reuse, remanufacturing and recycling of industrial waste and product after-use waste. Current legislation and regulations on waste do not always support the development of symbiotic relationships for companies. IE researcher and professionals need to work with policy makers to improve this situation to remove this legal barrier to IE applications in practice.

Clearly, the four areas in IE are closely related and should be integrated into IE applications. For example, developing IS exchanges supports the development of industrial ecosystems. Improving IM of industrial ecosystems requires continuous development of IS networks.

Governmental legislation and regulations need to be in line with IE development and applications. Through the integration of these four areas in their applications, IE contributes to environmental sustainability more profoundly.

It is the relativity of improvements that matters and has practical meanings, as each single step and each single symbiotic relationship establishment help move towards a higher level of a nearly closed-loop industrial ecosystem of material exchanges and energy cascading. A systemic perspective needs to be adopted to identify how individual symbiotic collaborations interact with each other, which gives the overall best effect as an industrial ecosystem and as an extended industrial ecosystem, with less interaction with the ecosystem of the Earth to reduce the impact.

References

Ayres, R. U. (1989). Industrial metabolism. In J. H. Ausubel & H. E. Sladovich (Eds.), *Technology and Environment* (pp. 23–49). Washington, DC: National Academies Press.

Ayres, R. U. (1994). Industrial metabolism: Theory and policy. In B. Allenby & D. J. Richards (Eds.), *The Greening of Industrial Ecosystems* (pp. 23–37). Washington, DC: National Academic Press.

Chertow, M. (2000). Industrial symbiosis: Literature and taxonomy. *Annual Review of Energy and the Environment, 25,* 313–317.

Côte, R., & Hall, J. (1995). Industrial parks as ecosystems. *Journal of Cleaner Production, 3*(1–2), 41–46.

Ehrenfeld, J. (2004). Industrial ecology: A new field or only a metaphor? *Journal of Cleaner Production, 12,* 825–831.

Fischer-Kowalski, M. (2003). On the history of industrial metabolism. In D. Bourg & S. Erkman (Eds.), *Perspectives on Industrial Ecology* (pp. 35–45).Sheffield: Greeleaf.

Frosch, R. A., & Gallopoulos, N. E. (1989). Strategies for manufacturing. *Scientific American, 261*(September), 144–152.

Gibbs, D., & Deutz, P. (2005). Implementing industrial ecology? Planning for eco-industrial parks in the USA, *Geoforum, 36,* 452–464.

Gui, L., Atasu, A., Ergun, O., & Toktay, L. B. (2016). Efficient implementation of collective extended producer responsibility legislation. *Management Science, 62*(4), 1098–1123.

Heeres, R. R., Vermulen, W. J. V., & de Walle, F. B. (2004). Eco-industrial park initiatives in the USA and the Netherlands: First lessons. *Journal of Cleaner Production, 12*, 985–995.

Jensen, P. D., Basson, L., Hellawell, E., Bailey, M. R., & Leach, M. (2011). Quantifying 'geographic proximity': Experiences from United Kingdom's National Industrial Symbiosis Programme. *Resources, Conservation and Recycling, 55*, 703–712.

Korhonen, J. (2002). Two paths to industrial ecology: Applying the product-based and geographical approaches. *Journal of Environmental Planning and Management, 45*, 39–57.

Leigh, M., & Li, X. (2015). Industrial ecology, industrial symbiosis and supply chain environmental sustainability: A case study of a large UK distributor. *Journal of Cleaner Production, 106*, 632–643.

Lewis, H. (2005). Defining product stewardship and sustainability in the Australian packaging industry. *Environmental Science & Policy, 8*, 45–55.

Lifset, R., Atalay, A., & Naoko, T. (2013). Extended producer responsibility. *Journal of Industrial Ecology, 17*, 162–166.

Lombardi, D. R., & Laybourn, P. (2012). Redefining industrial symbiosis. *Journal of Industrial Ecology, 16*, 28–37.

Lu, Y., Chen, B., Feng, K., & Hubacek, K. (2015). Ecological network analysis for carbon metabolism of eco-industrial parks: A case study of a typical eco-industrial park in Beijing. *Environmental Science and Technology, 49*, 7254–7264.

Malcolm, R., & Clift, R. (2002). Barriers to Industrial Ecology, The strange case of "The Tombesi Bypass". *Journal of Industrial Ecology, 6*, 4–7.

Mirata, M. (2004). Experiences from early stages of a national industrial symbiosis programme in the UK: Determinants and coordination challenges. *Journal of Cleaner Production, 12*, 967–983.

Park, J. M., Park, J. Y., & Park, H.-S. (2016). A review of the eco-industrial park development program in Korea: Progress and achievement in the first phase, 2005–2010. *Journal of Cleaner Production, 114*, 33–44.

Rogers, D. S., Rogers, Z. S., & Lembke, R. (2010). Creating value through product stewardship and take back. *Sustainability Accounting, Management and Policy Journal, 1*(2), 133–160.

Shi, H., Chertow, M., & Song, Y. (2010). Developing country experience with eco-industrial parks: A case study of the Tianjin Economic-Technological Development Area in China. *Journal of Cleaner Production, 18,* 191–199.

Tian, J., Shi, H., Chen, Y., & Chen, L. (2012). Assessment of industrial metabolisms of sulfur in a Chinese fine chemical industrial park. *Journal of Cleaner Production, 32,* 262–272.

Tibbs, H. (1992). Industrial ecology, an environmental agenda for industry. *Whole Earth Review,* winter, 4–19.

Velenturf, A. P. M. (2016). Promoting industrial symbiosis: Empirical observation of low-carbon innovations in the Humber region, UK. *Journal of Cleaner Production, 128,* 116–130.

Wassenaar, T. (2015). Reconsidering industrial metabolism: From analogy to denoting actuality. *Journal of Industrial Ecology, 19,* 715–727.

Xiang, W., & Ming, C. (2011). Implementing extended producer responsibility: Vehicle remanufacturing in China. *Journal of Cleaner Production, 19,* 680–686.

Zhang, Y., Qiao, Q., & Yao, Y. (2015). Study of eco-industrial park concept and connotation. *Applied Mechanics and Materials, 737,* 974–979.

4

Applications of Industrial Symbiosis

Abstract Industrial Symbiosis (IS), a study area within Industrial Ecology (IE), focuses on the knowledge web establishment of novel exchanges for synergies among companies to develop industrial ecosystems. Three types of IS applications have been explored in the literature: regional community-based IS, national IS programmes and eco-industrial parks (EIPs). These IS applications have offered valuable lessons. Critical success factors drawn from these practices are: an IS coordinating centre, economic and environmental gains in the vision, a large database of knowledge webs for potential symbiotic exchanges, early involvement of participating companies, and government investment at the start. IS applications need not be restricted by geographic proximity. Industrial clusters also need to be transformed into eco-industrial clusters. The transformation requires planned and facilitated IS and long-term vision.

Keywords Industrial symbiosis applications · Kalundborg industrial symbiosis (IS) · The UK national IS programme (NISP) Eco-Industrial parks (EIPs) · Geographic proximity · Eco-Industrial clusters (EICs)

4.1 Introduction

Industrial Symbiosis (IS) has been developed into a study area within the field of Industrial Ecology (IE). IS has its own elements of focus. IS explores ways to establish knowledge webs of novel material, energy and waste exchanges and business core processes to facilitate the development of networks of synergies within and across different companies to support the development of industrial ecosystems with high levels of nearly closed-loop material exchanges and efficiency of energy cascading. Novel exchanges are characterised by 'novel sourcing of required inputs and value-added destinations for non-product outputs', supported by improved business and technical processes and culture change (Lombardi and Laybourn 2012, pp. 31–32).

IS has developed through its applications, guided by IE principles. Typical IS applications can be classified into the following three categories:

1. Regional community-based IS development. The classic example is the IS Kalundborg practice in Denmark.
2. National-based IS programmes. The National IS programme in the UK is a representation.
3. Eco-industrial parks (EIPs). There are many EIPs for development around the world.

Following this introduction, the first three sections critically review each of the three IS application types respectively. In addition, there is a debate on the conditions that are essential to IS applications, including geographic proximity and diversity of industries. Section 4.5 addresses these conditions.

The critical review leads to lessons learned and also justification of some conditions for IS applications. The role of IS in achieving the ultimate goal of IE, which is to develop industrial ecosystems with high levels of closed-loop material exchanges and high efficiency of energy cascading, becomes clearer.

4.2 Regional Community-based Industrial Symbiosis Development - Kalundborg in Denmark

IS collaborations, characterised by novel sourcing of required inputs and value-added destinations for non-product outputs, have gradually progressed in Kalundborg, Denmark over the past forty-five years (Valentine 2016). The first collaboration took place between an oil refinery and a wallboard manufacturer in 1972 (Valentine 2016). The initiation was generated from social interactions that took place amongst Rotary Club members, a group of executives of local firms in Kalundborg discussing and sharing their key business challenges (Valentine 2016). It was long before the terminologies of IS, industrial ecosystem and IE were proposed over the period from 1989 to 1992 (Frosch and Gallopoulos 1989; Tibbs 1992). Unwanted natural gas, a by-product from the oil refinery process was transferred through a constructed pipeline to the wallboard manufacturer to be used to dry gypsum boards. It was a collaboration of creating 'a novel sourcing of required inputs and value-added destinations for non-product outputs' for the two companies involved while turning a by-product into an energy resource. Therefore, this collaboration is a symbiotic collaboration. Over four decades, more than 20 enterprises have been engaged in more than 30 symbiotic exchanges in the Kalundborg IS community (Branson 2016; Valentine 2016). This has been coupled with improved business and technical processes and business environmental culture in Kalundborg (Valentine 2016).

The IS community in Kalundborg has developed by involving the following types of companies across different industries: a coal-fired power plant, an oil refinery, a pharmaceutical plant, a wallboard manufacturer, a cement construction material manufacturer, a soil remediation company, pig farms, fish farms, local farms and others (Branson 2016; Ehrenfeld and Gertler 1997; Jacobsen 2006; Valentine 2016). It also includes neighbourhood residential households that receive waste heat as energy sources through the Kalundborg utility company (Ehrenfeld and Gertler 1997). The coal-fired power station alone has more than 10 symbiotic relationships with other local companies in Kalundborg, but many are energy and water-related exchanges. Besides the main production output of electricity, the coal-fired power station also generates waste heat, steam and other solid wastes, such as fly ash and sludge. The knowledge is critical and powerful to IS development, as well as opportunities to develop trust for collaboration. Knowing that fly ash can be used as a raw material for producing cement, scrubber sludge as fertiliser for farming, and gypsum in sludge for manufacturing plasterboards, made each initialisation of these IS relationships possible in the first place in Kalundborg. Table 4.1 presents some typical novel symbiosis exchanges within Kalundborg (Branson 2016; Ehrenfeld and Gertler 1997; Jacobsen 2006; Valentine 2016). The ideas of these novel exchanges can be fostered in similar types of companies in different locations worldwide, or help generate businesses initiatives.

During the first twenty years or so of the Kalundborg IS development, the executives of these enterprises were local residents. They were bound together through a social club gaining trust and providing opportunities for collaboration. At that time, collaborations were not specifically symbiotic collaborations. Further development of the IS network has benefited from the 'Environment Club' established in 1988, and the Kalundborg Symbiosis Centre in 1996. By then, many key decision-makers in these companies were no longer local residents. In the present day, the Kalundborg Symbiosis Centre not only acts as a centre to respond to information requests from individuals and researchers from outside of Kalundborg, but also plays a key role in hosting events and coordinating meetings and continues to connect executives from companies in Kalundborg and beyond.

The Kalundborg IS is a classic example of a gradually developed regional community of IS practice. Strictly speaking, the Kalundborg IS is not an EIP, even though some studies have described it so (Chertow 2000; Gibbs and Deutz 2007). Industrial parks, either eco-industrial parks or other types of industrial park, are purposefully built industrial areas with clearly defined geographic boundaries and under a park administration. The Kalundborg IS has been used as a case study by a number of authors to examine IE in practice or IS development (Branson 2016; Domenech and Davis 2011; Ehrenfeld 1997; Jacobsen

4 Applications of Industrial Symbiosis

Table 4.1 Typical novel exchanges of symbiotic collaborations within Kalundborg

Resource exchanged	From	To	Purpose	Transfer type
Natural gas	Oil refinery	Wallboard manufacturer	As an energy source burnt to dry gypsum boards	By-product to energy
Waste water	Oil refinery	Power station	Secondary use of water	Waste water to water resource
Cooling water	Oil refinery	Power station	Feeder water for boilers	Waste water to water resource
Sulphur	Oil refinery	Soil remediation company	Producing liquid fertiliser	Solid waste to input resource
Fly ash	Power station	Cement manufacturer	To produce cement	Solid waste to input resource
Heated cooling water	Power station	Fish farm	Warmer water throughout the year	Waste energy and water to energy and water resource
Salty cooling water	Power station	Fish farm	For increasing productivity	Water to water resource
Heat	Power station	Fish farm	Increasing water temperature for higher outputs	Waste energy to energy resource
	Power station	Residential households	Heating	Waste energy to energy resource
Steam	Power station	Oil refinery	For production process	Waste energy and water to energy and water resource
	Power station	Pharmaceutical manufacturer	For production process	Waste energy and water to energy and water resource
Scrubber sludge	Power station	Local farms	As fertilisers for farming	Solid waste to input resource
Yeast slurry	Pharmaceutical manufacturer	Pig farms	As stock food	Solid waste to input resource
Sludge (treated)	Pharmaceutical manufacturer	Local farms	As fertiliser	Solid waste to input resource

2006; Valentine 2016). In addition, the Kalundborg IS has been included as part of the IE or IS literature review by a number of authors (e.g. Chertow 2000; Chertow 2007; Dunn and Steinemann 1998; Lowe and Evans 1995; Tibbs 1992).

The Kalundborg IS is described as 'an organically-evolving, self-sustaining environmental collaboration in Denmark' (Valentine 2016, p. 65). Its development is characterised by a spontaneous initialisation, relying on social interactions of the local club and committee, followed by purposeful development facilitated by the Symbiosis Centre (Valentine 2016). This indicates that an organisation, either informal or formal, such as a club, committee or centre, is needed to facilitate interactions among decision-makers in different companies to provide opportunities to discuss and create symbiotic collaborations. This self-organising and self-sustaining feature were highlighted by a number of researchers (Valentine 2016; Chertow and Ehrenfeld 2012), and the transferability of the Kalundborg IS development is also questioned (Ehrenfeld 1997). Transferring a similar Kalundborg practice is unrealistic, undesirable and restricted by many conditions and factors. It is the underlying principle of initialising and developing business symbiotic collaborations through the aid of an organisation, such as a club, committee and centre, which should be transferable. The organisation facilitates and fosters the establishment and development of symbiotic relationships, either informally or formally.

The analysis of the Kalundborg IS in the literature focuses on IS exchanges of companies in the geographic area of Kalundborg. However, some Kalundborg companies have developed IS exchanges with companies outside of Kalundborg. These include, for example, an acid plant in Jutland, more than 200 miles away from Kalundborg, a power plant in Germany and a process company in the UK (Branson 2016; Ehrenfeld 1997). The acid plant in Jutland receives sulphur from the oil refinery in Kalundborg. The wallboard manufacturer in Kalundborg receives one-third of its gypsum supply from a German power station (Ehrenfeld 1997). 'Orimulsion ash' was sent to UK for processing (Branson 2016). This indicates that the Kalundborg IS practice has also established external symbiotic relationships, which is consistent with an extended systemic boundary principle in IS and IE.

The analysis of the Kalundborg IS practice in the literature generally only includes waste energy, waste water, industrial solid waste and by-products to be used as novel sourcing of required inputs for other companies. Resource inputs and product outputs have not been considered in analyses or illustrations of the Kalundborg IS practice by different researchers, including Chertow (2000), Domenech and Davis (2011), Ehrenfeld (1997), Jacobsen (2006) and Valentine (2016). For example, the resource inputs can be coal for the power station and raw materials for the pharmaceutical manufacturer, and product outputs can be plasterboard for the wallboard manufacturer and medicine for the pharmaceutical manufacturer. The illustration used in teaching slides by Cushman-Roisin (2013) included coal as an input and plasterboard as an output. The highlight is the collective set of energy, water, waste and by-product exchanges among companies to give mutual economic benefits, which also happened to turn waste to resource resulting in environmental benefits as a bonus (Heeres et al. 2004).

The development of a nearly closed-loop industrial ecosystem was not the intended focus for exploration of the Kalundborg IS practice in these studies. No doubt, these exchanges have reduced the intake of virgin materials from nature and waste and emissions to our biosphere. These symbiotic exchanges have increased the level of closed-loop material exchanges and efficiency of energy exchanges overall within the Kalundborg industrial ecosystem and beyond. The percentage and efficiency of these symbiotic exchanges of energy, waste and by-products over the total material and energy flows remain unknown and, therefore, these need to be explored and quantified. The change in the level of closed-loop material exchanges and energy cascading within this industrial ecosystem needs to be measured over time. Industrial metabolism (IM), an area within the field of IE, is concerned with the quantification of the rates and efficiency of these flows and their changes over time. Figure 4.1 illustrates some key IS exchanges of **solid waste only** within and beyond Kalundborg and also some key flows of materials inputs and product outputs to this industrial ecosystem, which have not been considered in the literature. Types of companies instead of company names, unlike other Kalundborg IS illustrations, have been used intentionally to capture the types of exchanges among companies directly.

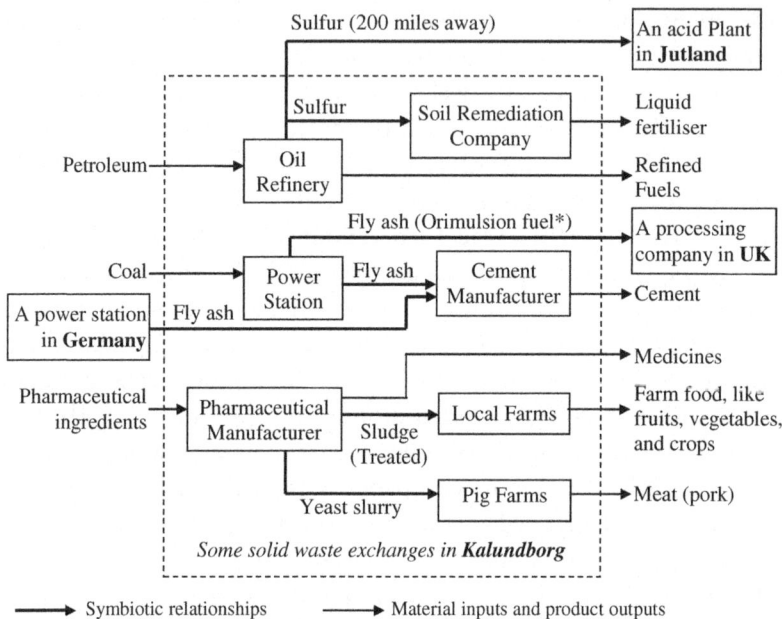

* Fly ash mentioned in other places is coal-fired fly ash. This symbiotic relationship ended when Orimulsion prices increased and the power station stopped using Orimulsion fuel.

Fig. 4.1 Some solid waste symbiotic exchanges and external material inputs and outputs excluding water and energy of some businesses in Kalundborg

The lessons learned from the Kalundborg IS development are as follows:

1. Every single IS exchange from industrial waste, by-products, water and energy to a resource is valuable to the development of an industrial ecosystem, and should be fostered.
2. A local club, committee or symbiosis centre, engaging with business executives and key decision-makers, is essential to provide connective opportunities, leading to symbiotic business collaborations.
3. A corporate environmental culture and pragmatic environmental mindset need to be developed.
4. Economic benefits must be present in any IS applications for long-term viability of synergies.

It is specified in a number of studies that economic gains are critical for these collaborations (Ehrenfeld 1997; Valentine 2016). The positive element is that there is an absence of conflict between the economic and environmental gains in these IS exchanges. The economic driven symbiotic collaboration brings environmental benefits simultaneously (Ehrenfeld 1997; Valentine 2016). The economic benefits of these exchanges of wastes and by-products were gained from saved landfill costs and cheaper input material costs (Jacobsen 2006). If a regulation does not impose disposal cost, for example, for sludge or fly ash, would these companies still pursue these collaborations? This may imply the importance of governmental legislation and regulations, as well as environmental consciousness and culture development in businesses to environmental sustainability development.

Lombardi and Laybourn (2012, p. 28) state that 'IS is not essentially localised waste and by-product exchanges'. The Kalundborg IS regional community development shows us one way to develop IS relationships. However, waste from after-use products of these businesses has not been considered. Further IS collaboration between industries, recycling companies and final consumers in extended system thinking brings a holistic view of the industrial ecosystem development.

4.3   UK National Industrial Symbiosis Programme

A national approach to IS practice has been adopted in the UK, although it began initially through pilot schemes in different regions. The Business Council for Sustainability Development-United Kingdom (BCSD-UK) initialised and coordinated the development of an IS network in the Humber area of the Yorkshire & the Humber region, North East England in 2000 (Mirata 2004), which grew further into other regions. The BCSD-UK officially launched a national IS programme (NISP) in 2003. The NISP had three pilot schemes in three regions: Scotland, West Midlands and Yorkshire & the Humber, led by Peter Laybourn who had a leadership role in the BCSD-UK (NISP 2017). The success of these pilot IS programmes led to investment from the

Department for Environment, Food and Rural Affairs (Defra), which enabled the NISP to roll out across all nine English regions in 2005 (NISP 2017). The nature of the operations in a company, the extent of peer pressure, the coordinating body and the approach of raising awareness of the IS programme in the regions, influenced the progress of the NISP in its early development (Mirata 2004). By February 2010, the NISP operated in twelve national regions (Jensen et al. 2011). The NISP continues developing its business network which includes more than 15,000 industrial members in early 2017 (NISP 2017), having increased from 13,000 in 2010 (Jensen et al. 2011). This substantial network allows easy identification of synergic exchanges, including energy, waste, water and logistics, which are mutually beneficial to companies (NISP 2017).

In 2005–2006, Peter Laybourn moved on to establish a consultancy company, named International Synergies Ltd, and took the responsibility of the NIPS coordination with him. The NISP coordinating team in the BSCD-UK and later in the International Synergies Ltd acted as the centre for developing different regional IS programmes and a NISP central database. The NISP centre facilitates information exchanges within and across different regional IS programmes and provides feedback regarding environmental sustainability to policymakers. The NISP centre performs analyses of material and energy flows to identify the possibilities of symbiotic relationships between companies, and also helps develop a more efficient and effective utility and waste management infrastructure for companies. The knowledge of the NISP team and their insight into specific local agendas are crucial to the success of these IS programmes and the NISP overall. The NISP has had a significant, positive impact on the UK economy and business environment (Jensen et al. 2011).

The NISP involves both regional and national government bodies and private companies. The NISP industrial members include micros, small and medium businesses (SMEs), as well as multinational corporations from every industrial sector (NISP 2017). The NISP follows the IS definition by Lombardi and Laybourn (2012) to focus on the establishment of knowledge networks as well as collaborations for novel exchanges including both physical resources and knowledge/information, without the restriction of geographic proximity. The information and knowledge exchanges facilitate materials and energy exchanges.

The NISP has become 'the largest national-level facilitated IS programme to date' in the world. The NISP's large network of businesses facilitates IS collaborations through a 'deliberate attempt' instead of 'self-organising' (Jensen et al. 2011, p. 703). The initial contact is normally made by a company to the NIPS team. NIPS practitioners are invited to make a company site visit to explore the problems of waste production and management in the company and gain a holistic view of the company's operational practice. What the company has and wants to achieve in relation to expertise, waste streams, by-products, and capacity of utilities are explored, recorded and entered into the NISP central database. Resource matches are identified and synergies are facilitated. After a completed and signed-off synergy, the social, environmental and economic outcomes are calculated and recorded in the NISP central database. The synergy outcome includes: 'amount of landfill diversion, reduction in virgin material use, reduction in CO_2 emissions, industrial water savings, hazardous waste elimination, jobs saved and/or created, cost savings, additional sales and any new private investment' (Jensen et al. 2011, p. 705). Best practices are also recorded in the database and fed into NISP workshops.

The physical symbiotic exchanges in the NISP include the following twenty categories by the end of the first five years of the NISP in 2007: coatings, WEEE (Waste Electrical and Electronic Equipment), infrastructure, glass, paper & cardboard, foodstuffs including oils, compost & soils, minerals, organic chemicals, wood products, composite packaging, misc. plastics, metals, ashes & slag, fuels, aqueous sludge, textiles, inorganic chemicals, rubber and hazardous wastes (Jensen et al. 2011, extracted from a table at p. 708). The knowledge and trust provided by the NISP coordinating centre are crucial to identify and facilitate a symbiotic relationship. The large database of the network developed by the NISP and the coordinating centre which by-passes trust building between companies have made the NISP successful in establishing synergies.

The IS programme in the NISP in the Humber area of the Yorkshire & the Humber region, North East England has its own uniqueness. Biomass waste resource has been used for power generation and fuel manufacturing (Velenturf 2016). Biomass waste resource partners have been developed in this area to increase business resource efficiency, lower business carbon emissions and increase business growth (Velenturf 2016). Five types of biomass waste resources have been explored: waste oils, agricultural feedstock, refuse-derived fuel, waste-wood fuel and waste oils and fats. The knowledge on the biomass waste-to-resource innovation is critical for the IS programme in this area and offers a unique type of symbiotic collaboration for the NISP in the UK. Developing this type of innovation partnership reveals two approaches for establishing synergies over time (Velenturf 2016). The first approach is to engage new partners from new industries during the innovation process, followed by limiting the number of resource partners once the innovation is realised (Velenturf 2016). The second is to expand innovative activities within the existing network during the innovation stage, followed by increasing the number of resource partners to avoid long-term dependences on existing partners (Velenturf 2016). Regardless of the approach adopted, new partners are always needed to the development of IS practice.

Lessons learned from the NISP are as follows:

1. The establishment of a knowledge web of different industrial wastes, by-products, water and energy is essential to a planned or facilitated IS programme.
2. The development of a large database containing potential participating companies and also successful synergies provides great opportunities for matching companies for IS synergies.
3. The involvement of IS consultancy companies and professionals by-passes the first stage of trust building required for developing a synergy.
4. Evaluations on completed synergies regarding economic, social and environmental benefits provide learning and encouragement for future synergy development.
5. Government support and investment at an early stage are crucial.

Despite the success and scale of NISP reported in some articles, the exploration of the NISP in the current academic literature is relatively limited, particularly compared to other IS practices. The majority of these publications exploring the NISP were produced by the NISP staff and associates. More research and exploration from independent researchers can provide a more objective perspective to the NISP development and further insight for future IS applications.

4.4 Eco-industrial Parks

The concept of EIP was introduced by a business development company, the Indigo Development, in the USA in 1993 (Cushman-Roisin 2013; Lowe 1997; Zhang et al. 2015). An EIP is described as *'a community of manufacturing and service businesses seeking enhanced environmental and economic performance through collaboration in managing environmental and service issues including energy, water, and materials. By working together, the community of businesses seeks a collective benefit that is greater than the sum of the individual benefits each company would realise if it optimises its individual performance only. The goal of an EIP is to improve the economic performance of the participating companies while minimising their environmental impact. Components of this approach include new or retrofitted design of park infrastructure and plants, pollution prevention, energy efficiency, and inter-company partnering. Through collaboration, this community of companies becomes an "industrial ecosystem"* (Lowe 1997, p. 58). The co-located feature of these companies in an EIP has been added in a follow-on explanation by Lowe (2001). EIPs were considered as 'concrete realizations' of the IS concept (Chertow 2000). EIPs were also perceived as one of the most important practice forms of IE theory (Zhang et al. 2015).

EIPs have been developed across different countries since 1993, including the USA, Canada, the Netherlands, Finland, China, South Korea and India. Table 4.2 summarises the key characteristics of EIPs in four selected countries. EIPs in the USA and the Netherlands represent practices in Western countries and EIPs in China and Korea characterise

Table 4.2 Characteristics of EIPs in the USA, the Netherlands, China and South Korea

Country (Since-)	Led by	Initialisation of EIPs		Characteristics of the development process of EIPs
		New or existing IPs	Vision	
USA (1994-)	Government agencies and park developers	Mainly newly developed EIPs, and some existing IPs	Job creation, Green intention	• Initialised by local and regional governments • Attempted to establish symbiotic collaborations after the start of running a park, proved problematic
The Netherlands (1994-)	Companies with government support	Mainly existing IPs and some newly developed EIPs	Both economic/environmental gains	• Initialised by business associations • Involvement of companies • Straightforward in establishing symbiotic collaboration
China (2001-)	Government and company involvement	Mainly existing IPs and a few newly developed EIPs	Cleaner production, Circular economy	• National Pilot EIPs • Accreditation of demonstration EIPs • Well-defined process from application to accreditation • Local government and foreign investment • Periodical evaluation • A coordinating centre within a park
South Korea (2006-)	Government and company involvement	Existing IPs	Reduction of pollution of existing industrial parks	• Pilot EIPs • Government financial support at initial stage • Diverse and large-scale industrial sectors • Regional EIP centres • Environmental culture development

practices in Eastern countries. The EIP literature presents a detailed exploration of EIPs in these four countries.

Led by the USA Environmental Protection Agency (EPA), a number of EIPs in the USA were initiated in 1994 and some opened in 1996 (Gibbs and Deutz 2005; Heeres et al. 2004; Zhang et al. 2015). The local and regional governmental agencies initiated these EIPS and the main purpose was job creation with the intention of being 'green' (Gibbs and Deutz 2005). However, companies that were relocated to an intended EIP commonly found that establishing a symbiotic collaboration with other companies in the park was extremely difficult, despite 'IS partnering' being a clear intention at the planning stage (Gibbs and Deutz 2005). Consolidated plans for matching potential companies for their material and energy exchanges seemed impossible at the planning stage. Material and waste exchanges did not occur during the initial years of running these intended EIP, owing two main challenges. The first was the security of waste information for sharing and the second was legislation around handling waste (Gibbs and Deutz 2005).

Despite the challenges of establishing symbiotic relationships following initiation, some positive outcomes were recognised by companies in these intended EIP parks. The positive outcomes included: improved communications with co-located fellow companies, established informal exchanges, and developed databases to facilitate future physical material exchanges (Gibbs and Deutz 2005). Companies confirmed that the networking opportunity provided within the park generated these positive outcomes. This is consistent with the experience in the Kalundborg IS development regarding informal interactions providing opportunities for symbiotic collaborations. It also implies that gradual change might be one feature of developing EIPs (Gibbs and Deutz 2005), under the current business environment and governmental legislation and regulations. Communication channels and trust have to be established prior to establishing an IS collaboration. Face-to-face interaction within a supportive environment is an effective way to establish communication channels and trust, and explore opportunities for symbiotic collaborations.

Interestingly, as the vision for these parks in the USA was to create jobs (Gibbs and Deutz 2005), it can be said that this objective has been achieved. Reducing impact on the environment was not directly set out in the vision for these intended EIPs in the USA (Gibbs and Deutz 2005). The interview notes indicated that some of these parks were not actually EIPs, but slightly 'greener' than other industrial parks by having 'greener' buildings and deploying 'cleaner' production technology and equipment. However, symbiotic collaborations of materials and energy exchanges, which are essential for IS practice and EIPs, were absent in these parks (Gibbs and Deutz 2005). Would it make any difference if the environmental gain was an equally important objective as job creation? Would establishing symbiotic relationships be more straightforward if consultation of participating companies had taken place prior to recruitment?

EIPs in the Netherlands was considered more successful than those in the USA, based on a comparison between three selected EIPs in each country (Heeres et al. 2004). All three USA EIPs were initiated in 1994, one existing IP and two newly developed EIPs; whilst the three Dutch EIPs were initiated in 1994, 1996 and 1998, and all were existing IPs. It was reasoned that Dutch EIPs involved participating companies at the setting-up stage, which made the establishment of symbiotic exchanges among companies more straightforward. It might be easier to transform existing IPs than build new EIPs, which requires the recruitment and relocation of existing companies to the site. The key features of the successful Dutch EIPs are as follows (Heeres et al. 2004):

1. Consideration of both environmental and economic gains in the vision of EIPs,
2. Initiation by a local association of entrepreneurs or employers on behalf of its member companies, acting as an anchor tenant in the park,
3. Involvement of direct stakeholders for decision-making, with consultation of agencies and educational institutions,
4. Financial involvement of participating companies in the development of EIPs, and
5. Utility sharing at the initial stage, leading to the development of other symbiotic exchanges.

The EIPs in the USA and the Netherlands employed different visions for development and different practices for the involvement of participating companies at the initial stage. Involving participating companies and employees at an early stage to ensure successful implementation of any management programmes is also confirmed in the OM literature, for example, Just-in-time (JIT) and Total Quality Management (TQM).

In China, the State Environmental Production Agency (SEPA), now the Ministry of Environmental Protection (MEP), started to promote EIP development in the late 1990s, and at that time there were more than 1500 industrial parks (IPs) in China (Zhang et al. 2010). The requirement to improve both the economic and environmental performance of companies, regions and the nation by the central Government drove the movement of transforming these IPs to EIPs (Tian et al. 2014; Zhang et al. 2015). The first national pilot EIP was approved by SEPA in 2001 (Tian et al. 2014). The transformation process is managed by the government agencies and the process is still on-going. The EIP transformation process starts an application by an IP, followed by planning, implementation and performance evaluation, which leads to accreditation if successful. The accredited parks in China are termed 'demonstration EIPs'. The government investment provides incentives for IPs to apply. There were twenty demonstration EIPs and sixty-one trailed EIPs in May 2013 (Tian et al. 2012) and these numbers continue to increase. The accredited EIPs are subject to periodical evaluation every three years to maintain standards (Tian et al. 2014). The demonstration EIPs across different cities in China, for example, Tianjin Economic-Technological Development Area (TEDA), Dalian Economic and Technological Development Zone (DETDZ) and Beijing EIPs (Shi et al. 2010) have been richly explored in the literature. The two distinctive characteristics of these demonstration EIPs in China, before and after transformation, are high-tech development and diverse industries. A significant improvement in both economic and environmental performance of the accredited EIPs was reported based on the evaluation of 17 sector-integrated demonstration EIPs in China (Tian et al. 2014). The following success factors supported the EIP practice in China:

1. A well-defined process of selection, development, evaluation and accreditation for transforming IPs to EIPs,
2. High-tech and diverse industries in the existing IPs,

3. Governmental investment at the start, and
4. IS centres within the park for facilitation.

A national EIP development programme of three phases over fifteen years has taken place in South Korea since 2006 to transform the existing IPs to EIPs (Park et al. 2008, 2016). Economic benefits provide the main motivation for businesses to participate. However, it was clear that the government aimed to reduce pollution caused by heavy industrial activities over the last few decades by transforming these IPs into EIPs (Park et al. 2008). The experience of the first phase of the South Korea EIP programme has confirmed the following success factors (Park et al. 2016).

1. Diverse industrial sector parks, involving large-scale companies in a park,
2. Government financial support at the initial stage,
3. Consistent and appropriate legal and regulatory systems for supporting and promoting EIP projects,
4. Willingness and workload of managers, as well as the capability and culture of an organisation, and
5. Regional EIP centres for facilitation.

Diverse and large-scale companies in a park provide more opportunities to match companies for symbiotic collaborations. Governmental financial support at the initial stage triggered the start of the transformation process. Both the 'Environment-friendly Industrial Structure Act' and the 'Industrial Cluster Development and Factory Establishment Act' were amended to support EIP development in Korea. However, legal issues hindered the development of resource sharing networks across the EIPs' boundaries and waste regulations remained a major issue for the first phase of the EIP programme (Park et al. 2016). The development of a flexible application and adaptation of waste regulations on a case-by-case basis through closer collaboration with regulatory bodies is still needed to ensure the success of the second phase of the EIP programme (Park et al. 2016). The regional EIP centres work with organisations to create new units to explore material recovery and circulation options and also to align with relevant personnel in companies to promote the EIP programme and secure jobs for workers. The regional EIP centres bring together different stakeholders from businesses, universities, research institutes and local governments through different forums to promote knowledge transfer, communication, information sharing and cooperation (Park et al. 2016).

The common features for transferring industrial parks to EIPs in China and South Korea are:

1. Government initiatives and investment,
2. Pilot projects to establish required databases, organisational infrastructure and developing strategies for wider participation, and
3. EIP centres to support identification of specific symbiotic collaborations among businesses.

It is confirmed that EIPs, compared with IPs, attract more business in China, Korea and the USA (Gibbs and Deutz 2005; Park et al. 2016; Tian et al. 2014). In addition, government investment for developing EIPs made an EIP a preferred option for businesses rather than an IP. The transformation of a large number of IPS to EIPs takes time and effort; but the successful transformation brings environmental, economic and social sustainability to nations and the world (Tian et al. 2014).

The limitation of analysis at an EIP level is that flows of the main products have not generally been considered. The system boundary of an EIP restricts consideration of the external inputs and outputs to the external environment of these EIPs (Gibbs and Deutz 2005; Lu et al. 2015). Extending a system boundary beyond an EIP can also add complexity. An analysis of carbon metabolism of a typical Beijing EIP revealed that the flow of input supplies and product demands outside an EIP was counted as the majority of the total material flow (Lu et al. 2015). The alternative way is to include the key flows of materials moving through an EIP. It also needs to be recognised that IS practice is more than an EIP programme and an approach considering a wider geographic area may be more viable in the long term (Gibbs and Deutz 2005).

Lessons learned from developing EIPs are as follows:

1. The involvement of participating companies at the planning stage and at the stage of recruiting fellow companies is essential,
2. An EIP centre for coordinating and providing opportunities for interactions among companies is critical, and
3. Both environmental and economic gains need to be included in the vision of an EIP.

EIP is only one type of industrial ecosystem and IS practice. Any symbiotic establishment between two facilities/companies contributes to the overall level of closed-loop material exchanges and energy cascading in an EIP. However, an EIP is still a long way from an industrial ecosystem with the high level of nearly closed-loop material exchanges and efficiency of energy cascading.

It is recognised that EIP development contributes to the development of industrial ecosystems. By analysing EIPs and their development, weak links and gaps in a geographic-based industrial ecosystem can be identified and improvements can be made accordingly. Further exploration regarding factors impacting on the success of EIP development and transformation as well as the relationship of an EIP with the external environment, both business environment and natural environment, is essential.

4.5 Conditions for IS Applications

This section explores the necessity of geographic proximity to IS applications, as well as the influence of industrial clusters and diversity of industries on IS applications.

4.5.1 Geographic Proximity

In 2000, Chertow emphasised that geographic proximity offered collaboration and synergetic possibilities among companies for physical exchanges of materials, water, energy, by-products and waste. Twelve years later, Lombardi and Laybourn (2012) argued that the time had changed for the necessity of geographic proximity as a condition for IS applications. In fact, symbiotic exchanges, including physical material exchanges, have never been restricted by geographic proximity. This has been demonstrated in many IS applications, such as the Kalundborg IS practice in recent years, the UK NISP and many EIPs.

Initially, the classic example of the Kalundborg IS practice and many EIPs seemed to provide confirmation that geographic proximity was necessary for IS applications. Geographic proximity appeared to have offered a great opportunity for co-located firms in Kalundborg to develop symbiotic relationships. It is certainly true that the majority of IS exchanges in Kalundborg have been undertaken within the geographic area of Kalundborg. Geographic proximity offered some convenience for executives of local firms in Kalundborg to have face-to-face contact to build trust and share business information and challenges at that time. However, it was simply not the case that geographic proximity led to IS collaborations or that IS collaborations required geography proximity. It was social interactions for trust building and information sharing, combined with pursuing business growth and economic benefits that led to the establishment of these IS exchanges in Kalundborg. Symbiotic exchanges have occurred both within and outside of Kalundborg. For example, for exchanges outside of Kalundborg, 'fly ash' was transported more than 100 miles to Aalborg, 'sulphur' more than 200 miles to Jutland, 'Orimulsion ash' to the UK for processing, and also some 'scrubber sludge' from a German power station (Branson 2016; Ehrenfeld and Gertler 1997).

The NISP certainly has not allowed geographic proximity to restrict its development in the UK. A dataset of 792 completed and signed-off synergies for the first five years of the programme from 2003 to 2007 confirmed the following: for half of these synergies, the distance material travelled was within 20.4 miles, three-quarters within 39.1 miles and the longest distance of 269 miles (Jensen et al. 2011). Only 13 per cent of these exchanges occurred within 5 miles, which can be considered as exchanges within a geographic proximity. Interestingly, the medium distance of synergies in TEDA is 21.2 miles, very similar to the one of the NISP. This indicates that geographic proximity is not a necessity for applying IS.

The development of EIPs certainly intends to foster synergies among companies within a park. The infrastructure of parks with geographic proximity provided opportunities to develop synergies among firms within parks (Boxi et al. 2015; Park et al. 2008; Shi et al. 2010). However, the majority of synergies in EIPs are cross-boundary synergies. For example, 59 per cent of synergies in the TEDA in China crossed the park's boundary (Shi et al. 2010; Jensen et al. 2011). In addition, the main flow was actually coming from external supplies and going out to external markets for a Beijing EIP (Lu et al. 2015). This indicates that a symbiotic exchange can occur beyond co-located companies within an EIP, even though a short-distanced material exchange has its advantages.

Accelerating an initialisation of IS collaboration requires the assistance of a knowledge web of possible symbiotic exchanges and an established network of companies to nourish novel ideas for exchanges of materials, water, energy, by-products and waste (Lombardi and Laybourn 2012; Mirata 2004). Extended system thinking and closed-loop flows of resources and energy, which are the key principles of IE, need to be considered in developing IS applications. Whenever possible, IS needs to minimise the distance travelled of materials, water, energy, by-products and waste for symbiotic exchanges, but IS applications should not be restricted by geographic proximity.

By appreciating that geographic proximity is not a restriction for IS applications, a holistic view of an extended system can be applied and IE principles can be considered when evaluating symbiotic exchanges and the overall impact of IS applications on environmental sustainability. It is explicitly stated in the literature that a collective way of an industrial ecosystem regarding its environmental and economic performance should be considered (Chertow 2000; Leigh and Li 2015). However, in practice, each synergy contributes to the overall industrial ecosystem's ability to absorb waste and by-products instead of emitting them to the natural environment.

Geographic proximity provides convenience for businesses to have face-to-face contacts that may lead to trust and information sharing, which are essential for firms to work collaboratively. This is not unique to the symbiotic relationship establishment, but to the business relationship development in general. Trust, effective information sharing (Mirata 2004), and social proximity are particularly critical to IS collaboration and innovation due to the involvement of diverse companies (Bansal and McKnight 2009; Letaifa and Rabeau 2013), but they are not necessarily achieved within a closely defined geographic area.

4.5.2 Industrial Clusters and Diversity of Industries

Lombardi and Laybourn (2012) distinguished from IS applications to industrial clusters, as the latter requires geographic proximity. They specified that IS should not be 'confused with agglomeration economies and industrial clusters', because 'IS engages diverse organisation in a network to foster eco-innovation and long-term culture change' (Lombardi and Laybourn 2012, p. 31). This raises two questions:

1. Is the diversity of companies essential for IS applications? If yes, at what level?
2. Can industrial clusters be transformed or evolved into eco-industrial clusters or EIPs?

Industrial clusters are 'geographic concentrations of interconnected companies and institutions in a particular field' (Porter 1998, p. 78), and the purpose of creating and developing industrial clusters is to gain a competitive advantage (Porter 1998). The characteristic of an industrial cluster is its specialised importance and competitive position in the industry, such as California Wine Cluster in the wine industry and Italian Leather Fashion Cluster in the fashion industry (Porter 1998). 'An array of linked industries' are co-located together to provide 'specialised' inputs and the infrastructure for producing specialised product outputs for a competitive market, along with institutions to support the maintenance and development of the specialisation through training and research (Porter 1998; Valdaliso et al. 2016). An industrial cluster does include diverse organisations or sectors. However, these organisations or sectors are interrelated by contributing to research, production, delivery and service of the same product families for competition in the market. For example, a specialised food manufacturing cluster contains a number of food manufacturers and their specialised growers for key ingredients, specialised tool and equipment makers, and research institutes. Another example is that 'in the late 1970s, the Basque papermaking cluster comprised about 30 manufacturing firms of paper and pulp (with an average size smaller than their foreign competitors), plus 50 firms of paper products, and a small number of manufacturers of machines and equipment for this industry' (Valdaliso et al. 2016, p. 72) and other organisations such as research institutions.

Industrial clusters have their diversity and heterogeneity of cluster knowledge embedded in the clustered firms, and network within and outside of a cluster (Valdaliso et al. 2016). The industrial diversity of a cluster might not have a right mix for developing IS initially, but it can be developed. In order to transform our industrial world to an eco-industrial world, we certainly need to transform industrial clusters as well as IPs. Sharing utility among companies in a cluster is certainly a good start. Collaborating with other companies within and outside of a cluster to develop IS exchanges is no different from cross-boundary collaborations, which occurred in the Kalundborg and some EIPs (Taddeo et al. 2012).

IPs have been proposed and constructed for different purposes compared to industrial clusters. IPs are designed to achieve the required economic growth by national or local governments. Constructing IPs was for the purpose of economic growth (Yu et al. 2014), rather than gaining competitiveness within an industry, like industrial clusters (Porter 1998). An IP normally contains different manufacturing companies in an industrial area, sharing the infrastructure of the park and the park management. An IP does not particularly locate companies with compatibility among their industrial activities to have industrial agglomeration effects (Yu et al. 2014). IPs can be either mixed-sector based or single-sector based. For example, a chemical manufacturing IP can have low value and high quantity of production but the products do not necessarily have the leading market position in the industry. The companies located in an IP, particularly for mixed sector industry parks, can be diverse, but they are not necessarily connected by a specialised industrial product family. IPs had been particularly popular in Eastern countries like China and Korea before this century. They served the purpose of increasing manufacturing capacities for economic growth. However, intensive manufacturing activities in IPs have generated heavy pollution, which has a significant negative impact on the environment. In order to reduce pollution, policies have been set and actions are taken to transform some of the IPs into EIPs and the transformation still continues (Parks et al. 2016; Zhang et al. 2010; Yu et al. 2014).

Transformation of both IPs and industrial clusters is necessary for developing high-level industrial ecosystems, which reduce intakes of virgin materials and disposal of waste and emissions. Research regarding the transformation of industrial clusters to eco-industrial clusters is very limited (Taddeo et al. 2012). Some explanations of IS do not actually help in transforming industrial clusters to eco-industrial clusters.

4.6 Summary

A knowledge web of different synergetic opportunities is extremely important for the IS network and relationship development. An informal and formal organisation acting as a coordinator to IS synergy development plays a vital role, particularly at the initiation stage. At this early stage, government support and investment are also crucial. The successful transformation of the current industrial world to an eco-industrial one offers long-term gains in all the three sustainability dimensions: economic, environmental and social. The private and public sectors, including research institutions and universities, need to work together to explore issues and identify solutions to contribute to research and the development of people in this field. A large database of relevant information of companies for potential IS collaborations regardless of regions and nations with the assistance of information technology, can accelerate the progress of transformation. Applying quantitative methods to evaluate social, environmental and economic benefits gained by a synergy provides evidence on the status of the closed-loop level of an industrial ecosystem and helps set the direction for future improvement. Early involvement of participating companies is particularly important to planned EIPs. Clear vision of environmental, social and economic gains can support the success of IS programmes. A corporate environmental culture and pragmatic environmental mindset need to be fostered in a community, region, nation and globe to support the transformation of our industrial world and society to an eco-industrial one.

IS applications predominantly employ a geographic-based approach to determine the boundary of an industrial ecosystem. The current applications of IS have come a long way from the theoretical debates on IS and its relationship with IE. EIPs and IS programmes are concrete applications of IS, even though they are still some way from the theoretical defined principles and goals of IS and IE.

Geographic proximity should not restrict IS applications and industrial clusters also need to be transformed into eco-industrial clusters, not just the transformation of IPs to EIPs. Every single effort leading to an IS exchange is valuable to the overall transformation of the industrial world to an eco-industrial world and eco-society.

IS applications in the literature also include specific settings of industries, such as cement manufacture (Hashimoto et al. 2010), solar photovoltaic cells production (Pearce 2008) and sugar production (Zhu et al. 2007). The details of these IS applications are outside of the space of this book.

References

Bansal, P., & McKnight, B. (2009). Looking forward, pushing back and peering sideways: Analyzing the sustainability of industrial symbiosis. *Journal of Supply Chain Management, 45*, 26–37.

Boix, M., Montastruc, L., Azzaro-Pantel, C., & Domenech, S. (2015). Optimization methods applied to the design of eco-industrial parks: A literature review. *Journal of Cleaner Production, 87*, 303–317.

Branson, R. (2016). Re-structuring Kalundborg: The reality of bilateral symbiosis and other insights. *Journal of Cleaner Production, 112*, 4344–4352.

Chertow, M. (2000). Industrial symbiosis: Literature and taxonomy. *Annual Review of Energy and the Environment, 25*, 313–317.

Chertow, M., & Ehrenfeld, J. (2012). Organizing self-organising system, towards a theory of Industrial Symbiosis. *Journal of Industrial Ecology, 16*(1), 13–27.

Cushman-Roisin, B. (2013). *The first tool of Industrial Ecology: Eco-Industrial Parks.* https://engineering.dartmouth.edu/~d30345d/courses/engs171/EIPs.pdf. Accessed February, 2017.

Domenech, T., & Davies, M. (2011). Structure and morphology of industrial symbiosis network: The case of Kalundborg. *Procedia Social and Behavioral Sciences, 10,* 79–89.

Dunn, B. C., & Steinemann, A. (1998). Industrial ecology for sustainable communities. *Journal of Environmental Planning and Management, 41*(6), 661–672.

Ehrenfeld, J. (1997). Industrial ecology: A framework and process design. *Journal of Cleaner Production, 5*(1–2), 87–95.

Ehrenfeld, J., & Gertler, N. (1997). Industrial Ecology in Practice: The Evolution of Interdependence at Kalundborg. *Journal of Industrial Ecology, 1* (1), 67–79.

Frosch, R. A., & Gallopoulos, N. E. (1989). Strategies for Manufacturing. *Scientific American, 261* (3), 144–152.

Gibbs, D., & Deutz, P. (2005). Implementing industrial ecology? *Planning for eco-industrial parks in the USA, Geoforum, 36,* 452–464.

Gibbs, D., & Deutz, P. (2007). Reflections on implementing industrial ecology through eco-industrial park development. *Journal of Cleaner Production, 15,* 1683–1695.

Grekova, K., Bremmers, H. J., Trienekens, J. H., Kemp, R. G. M., & Omta, S. W. F. (2014). Extending environmental management beyond the firm boundaries: An empirical study of Dutch food and beverage firms. *International Journal of Production Economics, 152,* 174–187.

Hashimoto, S., Fujita, T., Geng, Y., & Nagasawa, E. (2010). Realizing CO_2 emission reduction through industrial symbiosis: A cement production case study for Kawasaki. *Resources, Conservation and Recycling, 54* (10), 704–710.

Heeres, R. R., Vermulen, W. J. V., & de Walle, F. B. (2004). Eco-industrial park initiatives in the USA and the Netherlands: First lessons. *Journal of Cleaner Production, 12,* 985–995.

Jacobsen, N. B. (2006). Industrial symbiosis in Kalundborg, Denmark: A quantitative assessment of economic and environmental aspects. *Journal of Industrial Ecology, 10*(1–2), 239–255.

Jensen, P. D., Basson, L., Hellawell, E., Bailey, M. R., & Leach, M. (2011). Quantifying 'geographic proximity': Experiences from United Kingdom's National Industrial Symbiosis Programme. *Resources, Conservation and Recycling, 55*, 703–712.

Leigh, M., & Li, X. (2015). Industrial ecology, industrial symbiosis and supply chain environmental sustainability: A case study of a large UK distributor. *Journal of Cleaner Production, 106*, 632–643.

Letaifa, S. B., & Rabeau, Y. (2013). Too close to collaborate? How geographic proximity could impede entrepreneurship and innovation. *Journal of Business Research, 16*, 28–37.

Lombardi, D. R., & Laybourn, P. (2012). Redefining industrial symbiosis. *Journal of Industrial Ecology, 16*, 28–37.

Lowe, E. A. (1997). Creating by-product resource exchanges: Strategies for eco-industrial parks. *Journal of Cleaner Production, 5* (1–2), 57–65.

Lowe, E. (2001). *Eco-Industrial Handbook for Asian Developing Countries, prepared for the Environment Department, Asian Development Bank.* www.indigodev.com/Handbook.html.

Lowe, E., & Evans, L. (1995). Industrial ecology and industrial ecosystems. *Journal of Cleaner Production, 3*(1–2), 47–53.

Lu, Y., Chen, B., Feng, K., & Hubacek, K. (2015). Ecological network analysis for carbon metabolism of eco-industrial parks: A case study of a typical eco-industrial park in Beijing. *Environmental Science and Technology, 49*(12), 7254–7264.

Mirata, M. (2004). Experiences from early stages of a national industrial symbiosis programme in the UK: Determinants and coordination challenges. *Journal of Cleaner Production, 12*, 967–983.

NISP (National Industrial Symbiosis Programme). (2017). http://www.international-synergies.com/projects/national-industrial-symbiosis-programme/. Accessed 16 March, 2017.

Park, H.-S., Rene, E. R., Choi, S.-M., & Chiu, A. S. F. (2008). Strategies for sustainable development of industrial park in Ulsan, South Korea-From spontaneous evolution to systematic expansion of industrial symbiosis. *Journal of Environmental Management, 87*, 1–13.

Park, J. M., Park, J. Y., & Park, H.-S. (2016). A review of the eco-industrial park development program in Korea: Progress and achievement in the first phase, 2005–2010. *Journal of Cleaner Production, 114*, 33–44.

Pearce, J. M. (2008). Industrial symbiosis of very large-scale photovoltaic manufacturing. *Renewable Energy, 33* (5), 1101–1108.

Porter, M. E. (1998). Clusters and the new economics of competition. *Harvard Business Review, 76*(6), 77–90.

Shi, H., Chertow, M., & Song, Y. (2010). Developing country experience with eco-industrial parks: a case study of the Tianjin Economic-Technological Development Area in China. *Journal of Cleaner Production, 18,* 191–199.

Taddeo, R., Simboli, A., & Morgante, A. (2012). Implementing eco-industrial parks in existing clusters. Findings from a historical Italian chemical site. *Journal of Cleaner Production, 33,* 22–29.

Tian, J., Liu, W., Lai, B., Li, X., & Chen, L. (2014). Study of the performance of eco-industrial park development in China. *Journal of Cleaner Production, 64,* 486–494.

Tibbs, H. (1992). Industrial ecology, an environmental agenda for industry, *Whole Earth Review,* Winter, 4–19.

Valdaliso, J. M., Elola, A., & Orkestra, S. F. (2016). Do clusters follow the industry life cycle? Diversity of cluster evolution in old industrial regions. *Competitiveness Review, 26*(1), 66–86.

Valentine, S. V. (2016). Kalundborg Symbiosis: Fostering progressive innovation in environmental networks. *Journal of Cleaner Production, 118,* 65–77.

Velenturf, A. P. M. (2016). Promoting industrial symbiosis: Empirical observation of low-carbon innovations in the Humber region, UK. *Journal of Cleaner Production, 128,* 116–130.

Yu, C., Dijkema, G., & de Jong, M. (2014). What makes eco-transformation of industrial parks take off in China? *Journal of Industrial Ecology, 19*(3), 441–456.

Zhang, L., Yuan, Z., Bi, J., Zhang, B., & Liu, B. (2010). Eco-industrial parks: national pilot practices in China. *Journal of Cleaner Production, 18* (5), 504–509.

Zhang, Y., Qiao, Q., & Yao, Y. (2015). Study of Eco-Industrial Park Concept and Connotation. *Applied Mechanics and Materials, 737,* 974–979.

Zhu, Q., Lowe, E. A., Wei, Y.-A., & Barnes, D. (2007). Industrial Symbiosis in China: A Case Study of the Guitang Group. *Journal of Industrial Ecology, 11* (1), 31–42.

5

Life Cycle Thinking and Analysis, Design for Environment, and Industrial Ecology Frameworks

Abstract This chapter explores the product life cycle, life cycle analytical tools and design for the environment (DfE) methodology. The product life cycle from an operations management (OM) perspective includes material acquisition, manufacturing, distribution, use and after-use. DfE explores eco-design options at each stage of this product life cycle to proactively reduce the impact of industrial activities on the environment. This chapter also presents two Industrial Ecology (IE) frameworks, one at a factor level and the other one at a supply chain level. These two frameworks illustrate the importance of integration and collaboration among different parts and parties within and across industrial ecosystems to increase levels of closed-loop material, energy and waste flows, which reduce their interaction with natural systems, hence the reduced impact.

Keywords Life cycle thinking/analysis · Design for environment (DfE) Industrial Ecology (IE) frameworks

5.1 Introduction

Industrial Ecology (IE) has embraced different methodologies, approaches and analytical tools to develop industrial ecosystems with high level of nearly closed-loop material exchanges and efficiency of energy cascading. Approaches in Industrial Symbiosis (IS) and industrial metabolism (IM), which are also IE approaches, have been explored in previous chapters. This chapter addresses life cycle thinking/analysis and design for environment (DfE), which also assists the achievement of the IE goal. In addition, IE frameworks which can be used to guide the development of closed-loop industrial ecosystems are also presented in this chapter.

5.2 Life Cycle Thinking/Analysis

There are two different perspectives associated with the concept of the product life cycle: (1) a marketing perspective and (2) an operations management (OM) perspective including supply chain management (SCM). From the marketing perspective, the product life cycle consists of four basic phases: (product) introduction, growth, maturity and decline, in relation to general sales performance of a product or product family in the market, following product development and launch. This generates research and applications in product life cycle management (LCM), which integrates new product development, project management and product data management to achieve sustainable patterns of product consumption and production (Gmelin and Seuring 2014; PwC 2010). However, this book focuses on the product life cycle from the OM perspective, which supports the development of nearly closed-loop industrial ecosystems and is explored below.

From the OM perspective, a product's life cycle starts from resource acquisition and goes through manufacturing, distribution, use and after-use (Fig. 5.1). Inputs and outputs of each stage, options for decision-making, strategies and policies, as well as impacts on environment, economy and society can be explored along this product life cycle.

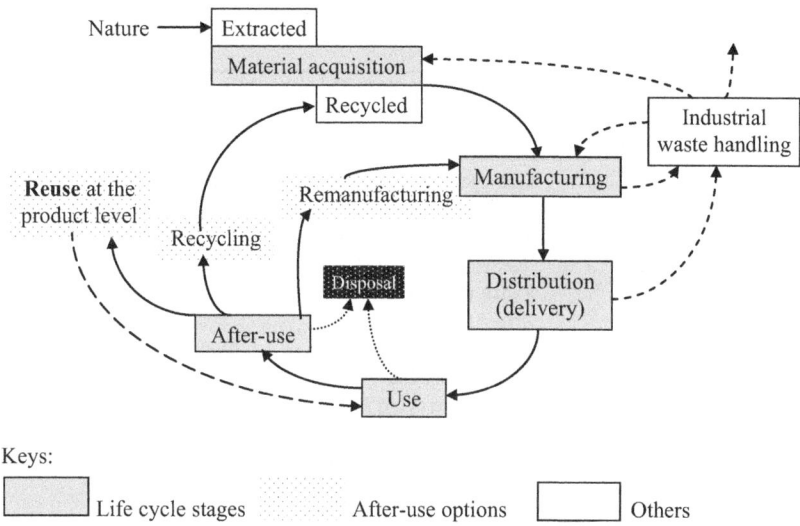

Fig. 5.1 Product life cycle with consideration of product after-use options

In this product life cycle, after-use options of reuse, remanufacturing and recycling are important for developing high levels of closed-loop material exchanges and efficiency of energy cascading, and therefore need to be fully explored before the final disposal is considered (Fig. 5.1). Besides 'waste' generated at the after-use stage, there is 'waste' generated at other stages in the product life cycle, such as industrial 'waste' from manufacturing and distribution. Industrial waste has been explored within the study area of Industrial Symbiosis (IS) in previous chapters. Figure 5.1 focuses on the main product life cycle and after-use options and does not intend to include all possibilities associated with each stage of a product life cycle.

Life cycle thinking/analysis has formed the basis for the development of various life cycle analytical tools and applications in the literature, with great divergence (Guinée et al. 2011). Terms used for life cycle analytical tools and their abbreviations are often inconsistent and do not always reflect the purpose which a tool was intended to achieve accurately, as they were proposed independently along their developments. Table 5.1 presents several life cycle analytical tools drawn from

Table 5.1 Life cycle analysis tools in the literature and some suggested terms and abbreviations

Aspect for application of life cycle thinking/analysis	Life cycle analytical tools Current terms (abbreviations)	Life cycle analytical tools Suggested terms (abbreviations)
Environment	Life cycle assessment (LCA[a]) • Life cycle inventory assessment (LCI) • Life cycle impact assessment (LCIA)	Life cycle environmental analysis (LCEnA)
Cost/Economic	Life cycle cost analysis (LCC[b] or LCCA)	Life cycle economic analysis (LCEcA)
Social	Social Life cycle analysis (SLCA)[b]	Life cycle social analysis (LCSoA)
Sustainability	Life cycle sustainability analysis (LCSA)[b]	(same)
Risk	Life cycle risk analysis (LCRA)	(same)

[a]Abbreviated as LCA in the ISO 14040 document.
[b]LCC, SLCA and LCSA were all mentioned in Guinée et al. (2011). SLCA can also stand for *Streamlined LCA* (Graedel and Allenby 1995, 2003).

the literature and also suggestions for new terms and abbreviations to the three sustainability life cycle analytical tools.

Life cycle assessment (LCA) was developed to evaluate and assess environmental impacts along the product life cycle by the International Organizations of Standardization (ISO 14040 2006). In the ISO 14040 (2006), LCA is defined as 'compilation and evaluation of the inputs, outputs, and the potential environmental impacts of a product system throughout its life cycle' (p. 2). LCA is also defined as 'a technique for cataloguing and assessing the impact of the resource use and environmental releases at each stage of the product life cycle (Lifset 2000, p. 404). LCA in the ISO1 14040 contains specific procedures and methods. The LCA procedure includes setting the goal and scope for assessment, quantifying inputs and outputs (inventory) for each stage of the product life cycle, assessing environmental impacts of a product along with its life cycle, and interpreting results (ISO 14040 2006).

In the ISO 14040 (2006), 'life cycle' is defined as 'consecutive and interlinked stages of a product system, from raw material acquisition or generation from natural resources to final disposal' (p. 2). In this life cycle definition by ISO, the product after-use stage is not specified. However, in the example of a product system used for LCA (ISO 14040 2006, p. 10), after-use options and waste treatment have been included. IE strongly considers product after-use options and the impact of product and process design on these options, as well as recycling and transformation of industrial waste at manufacturing and distribution stages. This consideration in IE has been reflected by the product life cycle illustrated in Fig. 5.1.

LCA is concerned with environmental aspects. However, the term 'life cycle assessment' does not adequately reflect this purpose, as the word 'environmental' is not present. The term 'life cycle environmental analysis (LCEnA)' would more accurately reflect a life cycle analysis for environmental aspects (Table 5.1). The same is applied to life cycle analysis terminologies for social and economic dimensions. Alternatives have been suggested in Table 5.1, such as LCEcA represents life cycle economic analysis and LCSoA life cycle social analysis.

Life cycle sustainability analysis (LCSA) includes life cycle analytical tools for all the three sustainability dimensions and is regarded as the future direction for life cycle thinking/analysis (Guinée et al. 2011). LCSA integrates LCEnA, LCEcA and LCSoA. However, LCSA dramatically increases the complexity and scale of a life cycle analysis. In practice, life cycle analysis in individual sustainability dimensions needs to be applied first before integrating them. Besides, life cycle analysis has been applied to other aspects, such as risk - life cycle risk analysis (LCRA), which can be further incorporated within any dimension in LCSA.

Figure 5.1 does not include other recovery options after all three after-use options have been considered prior to disposal. Other recovery options include anaerobic digestion, incineration, gasification and pyrolysis which produce energy and materials from wastes (Department for Environment 2014; Leigh and Li 2015). Consideration of other recovery options can further reduce the quantity of disposal. By totally avoiding disposal, the product life cycle becomes cradle to cradle

(C2C), rather than cradle to grave (C2G), as disposal represents the grave stage. Further reading on C2C can refer to Silvestre et al. (2014), Toxopeus et al. (2015) and Bjørn and Hauschild (2012).

Previously, a product life cycle was considered as C2G (Lifset 2000). However, life cycle thinking/analysis in IE takes the C2C perspective, which prioritises pollution prevention approaches, closed-loop material exchanges and efficient energy cascading. The life cycle analysis has been further incorporated with eco-design options to develop the DfE methodology in IE, which is explored in the next section.

5.3 Design for Environment

'The defining principle of DfE is the integration of environmental considerations at the inception of the product and process design phase' (Jackson et al. 2016, p. 144). DfE is 'a way to systematically consider design performance with respect to environmental, health and safety objectives over the full product or process life cycle' (Airbus ACADEMY 2017). The first definition needs to emphasise the role of the product life cycle in DfE. The second definition widens the focus of DfE by including health and safety objectives and also needs to specify the consideration of eco-design options, instead of design performance. In this book, DfE is regarded as a methodology that explores different eco-design options through implementing a set of approaches and strategies along each stage of the product life cycle from material acquisition, manufacturing/production, distribution, use and after-use. Table 5.2 presents DfE life cycle stages and associated approaches and strategies.

Approaches and strategies in DfE are all pollution prevention approaches and strategies as they explore product eco-design options for the reduction of pollution and emission and the increased level of closed-loop material exchange and efficiency of energy cascading. DfE approaches and strategies have been developed along the five stages of the product life cycle. All DfE approaches and strategies are important and need to be integrated into their applications to have a profound impact on environmental sustainability. However, this book focuses on

Table 5.2 Design for environment (DfE) and associated approaches and strategies

DfE—product life cycle stages	DfE approaches/strategies
Eco-design for *material selection*	• Raw material deletion (avoiding) • Renewable resources use (increasing) • Use recycled materials (increasing) • Recyclability (increasing) • Energy embedment (decreasing) • Hazard rating (reducing to zero) • Degradability (increasing) • Dematerialisation (increasing)

Eco-design for *manufacturing production process*	Product design with consideration of production process design, deployment and upgrading through the following:
	• Process life cycle (improving efficiency and prolonging life-span)
	  – Equipment life cycle
	  – Machine tool life cycle
	• Production process efficiency of energy consumption (increasing)
	• Industrial pollution and waste emission (reducing by applying industrial symbiosis (IS))
	• Product stewardship (increasing responsibility of all parties along a product life cycle in reduction of a product's impact on the environment)

Eco-design for *distribution and delivery*	• Design for packaging
	  – No packaging
	  – Minimal packaging
	  – Degradable, reusable, returnable, refillable packaging
	• Design for transportation (reduced transportation cost) (normally working with design for packaging)

Eco-design for *use*	• Design for reduced residue generation during product usage • Design for reduced energy consumption during product usage • Design for eco-efficiency of servicing

Eco-design for *after-use*	• Design for reuse
	• Design for remanufacture (DfRem)
	  – Design for remanufacture process
	  – Design for disassembly
	  – Design for upgrading
	• Design for recycling

approaches of product eco-design options for product after-use, as they are particularly relevant to the development of nearly closed-loop industrial ecosystems, through turning waste to resources.

Eco-design for after-use in DfE contains three key approaches: design for reuse, design for remanufacture (DfRem) and design for recycling (DfRec). Design for reuse can be either at the product level or part level. 'Design for reuse' at the part level is a critical element for 'DfRem' (Jayaraman 2006). Therefore 'design for reuse' at the part level is included in DfRem for discussion, and 'design for reuse' in this book is concerned with eco-design for reuse at the product level.

Approaches of eco-design for after-use directly impact on the efficiency and effectiveness of end-of-life approaches of reuse, remanufacturing and recycling. Most importantly, DfE upgrades end-of-life (or end-of-pipe) approaches to preventative integrated end-of-life approaches including 'reuse-after-design for reuse' (at the product level), 'remanufacturing-after-DfRem' (including reuse of parts) and 'recycling-after-DfRec' (Table 5.3). If these eco-design after-use approaches are effectively considered in product and process designs, they dramatically improve the efficiency and effectiveness of reuse, remanufacturing and recycling activities. This leads increased levels of nearly closed-loop material exchanges and efficiency of energy cascading within and across industrial ecosystems.

Reuse at the product level deploys different processes from reuse at the part level for remanufacturing and is commonly carried out by independent manufacturers through checking, cleaning, repairing, refurbishing and reconditioning of products for resale. Products are completed functioning entities at different levels, such as a car and its engine which can both be considered as products. An engine is part of a car, but has its own independent function to perform. Parts are those components that cannot perform an intended function independently and need to be assembled together with other parts to become a unit to perform an intended function. As repairing or reconditioning of a product for resale is not usually the main concern for original equipment manufacturers (OEMs), direct design considerations for reuse at the product level are limited. However, OEMs need to consider ease of repair/replacement of a part for after-sale service, or possibly

Table 5.3 Upgraded end-of-life pollution control approaches after integrating with 'design for after-use' pollution prevention approaches in DfE

Upgraded end-of-life pollution control approaches	Pollution prevention approaches	Description
Reuse-after-design for reuse at the product/unit level	Design for reuse (at the product/unit level)	Design considers ease of the following: checking, cleaning, repairing, refurbishing of the whole item. Reuse-after-design for reuse at the part level makes reuse at the product level more efficient functionally and economically and easier to carry out
Remanufacturing-after-design for remanufacturing (DfRem) (including design for reuse at the part level)	Design for remanufacture (DfRem), including part reuse	Design considers ease of the following: dissembling, sorting, cleaning, refurbishing of dissembled parts (reuse of parts) and the continuity of parts to be used and their compatibility of reassembly with new parts
Recycling-after-design for recycling (DfRec)	Design for recycling (DfRec)	Design considers the recyclability, which is the possibility of turning waste into new substances or materials, including recycling industrial waste and product or manufacturing facility after use

remanufacturing options. Therefore, eco-design for reuse at the part level benefits reuse at the product level.

The goal of DfRem is to enhance the possibility of remanufacturing—remanufacturability and DfRem are required to consider each step of the remanufacturing process at product and process design (Hatcher et al. 2011). Remanufacturing is different from repair and reconditioning at the product level. Unlike repair and reconditioning of a product, the original product identity is no longer retained after remanufacturing (Haynsworth and Lyons 1987). Remanufacturing restores worn-out products to like-new condition, through dissembling, sorting, cleaning and reassembling, by adding new parts if necessary, in a factory environment to 'produce a unit fully equivalent and sometimes superior in performance and expected lifetime to the original new product' (Haynsworth and Lyons 1987, p. 24).

Remanufacturing normally relies on 'remanufacturing-oriented designs' or DfRem, which permits OEMs to use and access end-life products to disassemble them for parts. DfRem includes a series of design activities, design for core collection (collection of discarded products with durable and quality parts for remanufacturing), design for dissembling, design for multiple lifecycles, design for upgrading and design for evaluation (Hatcher et al. 2011). As remanufacturing has both economic and environmental benefits to OEMs, independent remanufacturers, and consumers, as well as social benefits to societies, it has become a growing business. For example, in the car industry, remanufactured alternators and starters have more than 90 per cent of the replacement market and are produced at a fraction of the original cost (Bernard 2011). They require approximately 14 per cent of the energy and 12 per cent of the material needed to produce entirely new ones (Bernard 2011). 'Remanufacturing benefits society and the industrial world by reduced post-consumption waste, lower energy and raw material consumptions, and lower prices for replacement products' (Bernard 2011, p. 350).

DfRec needs to be incorporated with DfRem to allow worn-out parts that cannot be entered into the remanufacturing process to be recycled, rather than disposed. Recycling is an activity that processes waste into a new substance or material, like paper or plastics recycling. Increasing recyclability of selected materials through DfE enhances the result of recycling after-use products. Recycled materials can be redirected to any industries.

Approaches in DfE have both technical and managerial aspects and require inputs and integration of the two to be effective in practice. Some technical aspects of these approaches have been considered in material engineering, product and process development and design,

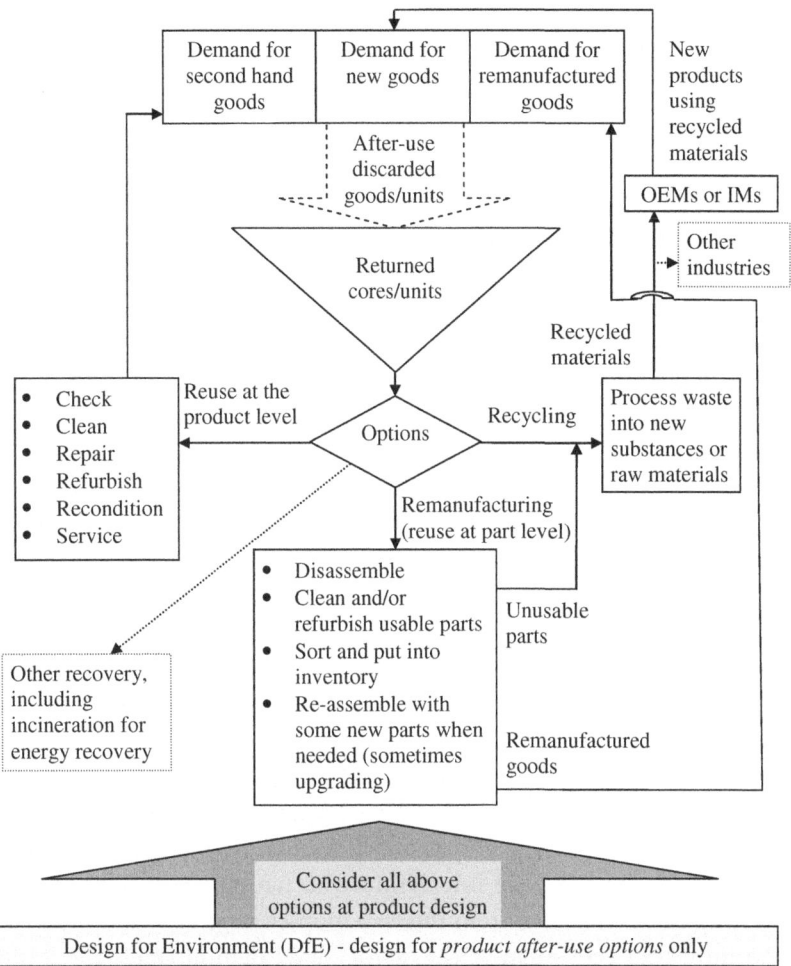

Fig. 5.2 Design for after-use options and contribution of DfE to closed-loop industrial ecosystems

and environmental sciences and management. More awareness and consideration of DfE need to be given to management disciplines, particularly OM, as OM focuses on continuous improvement of processes to deliver goods and services to customers. OM considers design, management and continuous improvement of processes to support the fulfilment of technical aspects to achieve operations objectives, and this is also relevant to the implementation of approaches and strategies of DfE. OM needs to consider the integration of the DfE methodology to explore eco-design options at the product and process design phase. A managerial process with consideration of three product after-use options supported by DfE is illustrated in Fig. 5.2.

DfE is particularly relevant to material- and product-based industrial ecosystems (for further explanations on types of industrial ecosystems, see the methods determining boundaries of industrial ecosystems in Sect. 3.2, Chap. 3). DfE helps achieve increased levels of closed-loop material exchanges and efficiency of energy cascading, particularly for material- and product-based industrial ecosystems.

5.4 Industrial Ecology Conceptual Frameworks

The key features of IE are integration and collaboration within and across industrial ecosystems to reduce the degree of interaction with natural ecosystems, which leads to higher levels of nearly closed-loop industrial ecosystems on different scales. The scale of an industrial ecosystem is subject to its boundary which varies according to study purpose. This section considers two scales of industrial ecosystems, a factory and a supply chain. IE frameworks on these two scales explore new paths of materials, energy and waste flows within an industrial ecosystem to reduce the degree of interaction with natural ecosystems through increasing industrial ecosystem internal loop-closing.

An industrial ecosystem at the factory level demonstrates the importance of integration among different parts within a small-scale industrial ecosystem (Fig. 5.3), normally within the same company. Figure 5.3 considers the integration of operations, the building where operation processes are located and supportive facilities including offices to operations, in order to produce goods and services. Figure 5.3 is developed based on the conceptual model of IE at the factory level by Despeisse et al. (2012), but in a very different representation.

For example, the amount of scrap generated by operations processes can be normally reduced by improving planning skills and deploying

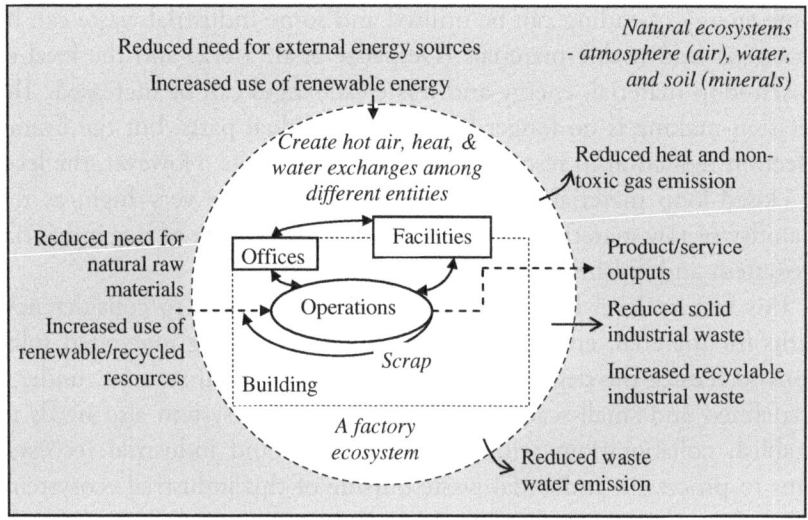

Fig. 5.3 A conceptual IE framework at a factory level

advanced technology. However, IE considers the overall impact of a system through the integration of operations processes with facilities, product and process design functions, and/or green technology. The integration of different parts in a factory allows the ease of reuse of scrap to reduce waste going out of this factory industrial ecosystem. Circulations of hot air, heat and water including waste water can be re-considered to create more exchanges and increased efficiency of energy cascading within the system. The overall impact of integration of different parts in this factory industrial ecosystem can be:

1. reduced need of natural raw materials, and
2. reduced waste and emissions.

With the aid of DfE, consideration can also be given to the use of more recycled materials, renewable natural materials and renewable energy in product and process designs along the product life cycle. This particularly focuses on potential at the production stage to further develop this industrial ecosystem. By integrating different parts in a factory, some energy cascading can be utilised and some industrial waste can be processed into usable materials (Despeisse et al. 2012) and the level of closed-loop material, energy and waste exchanges can be increased. The decision-making is no longer based on individual parts, but the overall effect on reduction in resource inputs and emissions. However, the level of closed-loop materials exchanges is unlikely to be very high, as the majority of raw materials are still coming from outside of this industrial ecosystem and finished products going out of it.

This factory level IE framework enables companies to consider new paths for material, energy and waste flows by seeking integrated solutions to reduce the degree of interaction with natural systems, under a predefined and small-scale system boundary. This system also needs to establish collaborations with other industries and industrial (eco)systems to process its industrial waste outside of this industrial ecosystem, either through a waste processor or an industrial partner from the same industry or across different industries. Collaborations among different industrial partners have been explored in Industrial Symbiosis (IS) in Chaps. 3 and 4. This section focuses on collaborations among different parties within a supply chain industrial ecosystem with the aid of the recycling industry.

A supply chain industrial ecosystem conceptual framework illustrates collaborations among different parties along a supply chain (Fig. 5.4), connected further by waste collectors/processors. This IE framework is a further development based on the conceptual framework for IE and IS implementation developed by Leigh and Li (2015), but in a very different representation. Connections of this industrial ecosystem with other industries and supply chains, either through waste collectors/processors or consumers/end-users have also been illustrated in this supply chain IE framework.

Different material flows within this industrial ecosystem can be created through waste collectors/processors and consumers/end-users. Even though this IE framework focuses on a single supply chain network environment, collaborations between different industries occur through waste collectors/processors and consumers/end-users. Core collection for remanufacturing is also illustrated through the link between consumers/end-users and manufacturing companies. Extended producer

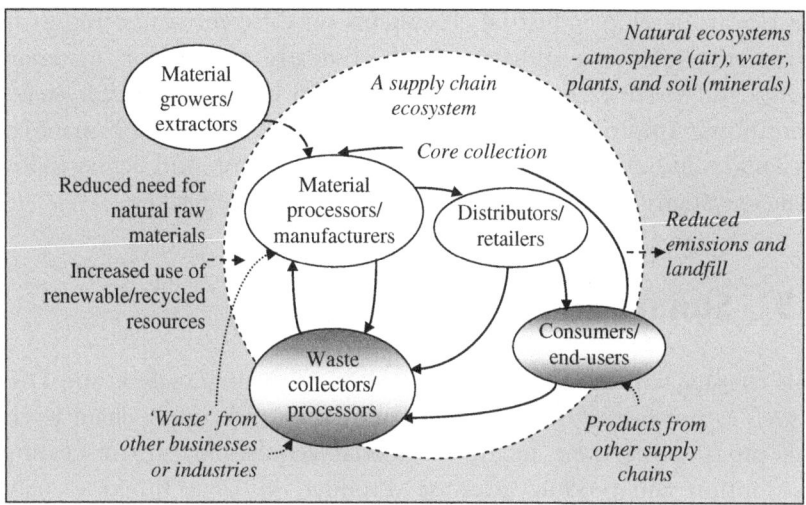

Fig. 5.4 A conceptual IE framework at a supply chain level

responsibility (EPR) can facilitate the establishment of this linkage (Lifset et al. 2013; Gui et al. 2016; Xiang and Ming 2011). The supply chain industrial ecosystem considers material extractors/growers as outsiders, and raw natural resources from them should be dramatically reduced through increasing the use of recycled materials, and also feedstock from other industries (important but not explicitly considered in this supply chain industrial ecosystem). Again, DfE supports the development of this supply chain industrial ecosystem along the product life cycle from material acquisition to product after-use to close material loops. New energy flows can also be created with collaborations for new energy cascading opportunities within this industrial ecosystem as well as the increased use of renewable energy from outside of this industrial ecosystem. By increasing collaborations, this supply chain industrial ecosystem as a whole requires less quantity of raw material from natural systems and emits less waste to natural systems.

Integrations and collaborations can be demonstrated by potential new paths of materials, energy and waste flows to increase levels of closed-loops within an industrial ecosystem, leading to a reduced degree of interaction with natural systems. IE frameworks can guide the practice to develop industrial ecosystems on different scales. Industrial ecosystems require a sufficient level of nearly closed-loop materials, energy and waste exchanges to be claimed to be industrial ecosystems. Continuous improvement in the level of nearly closed-loop materials exchanges and efficiency of energy cascading within and across industrial ecosystems by applying extended system thinking is key.

5.5 Summary

This chapter has explored product life cycle thinking/analysis and DfE, as well as presenting IE frameworks at factory and supply chain levels. The product life cycle includes material acquisition, manufacturing, distribution and product after-use. Product life cycle thinking/analysis supports a variety of life cycle analytical tools, including life cycle assessment (LCA), life cycle costing (LCC) and life cycle sustainability analysis (LCSA). However, the terminologies and abbreviations of these life cycle analytical tools are inconsistently used. Therefore, life cycle environmental analysis (LCEnA), life cycle economic analysis (LCEcA) and life cycle social analysis (LCSoA) have been proposed to address life cycle analysis for the three sustainability dimensions, environmental, economic and social. LCSA, which contains all three dimensions of sustainability analysis, dramatically increases the complexity of a life cycle analysis. DfE is based on the product life cycle analysis to explore eco-design options at each stage along the product life cycle. This provides a proactive methodology to incorporate product reuse at the product level, remanufacturing (including reuse at the part level) and recycling at the product and process design phase of new product development.

This chapter also presents two IE frameworks, one at a factory level and one at a supply chain level. The IE framework at a factory level illustrates the importance of integration of different parts in a factory environment. With the integration, decision-making no longer considers each part in isolation, but the system as a whole by creating new flows of materials, energy and waste, which reduce the degree of interaction with natural systems. The IE framework at a supply chain level emphasises the importance of collaborations among different parties along a supply chain and also waste collectors and processors. The material supply role of material growers and extractors has been dramatically reduced when a supply chain industrial ecosystem has increased the use of recycled materials and feedstock from other industries within the industrial ecosystem or across different industrial ecosystems. The important role of waste collectors/processors and consumers/end-users to the development of higher levels of nearly closed-loop industry ecosystems at a supply chain level is emphasised.

The closed-loop thinking in these two IE frameworks needs to be embedded in OM to support the paradigm shift in the representation of linear processes to closed-loop considerations of producing and delivering goods and services. IE needs to become part of the Dominant Social Paradigm (DSP) to influence decisions in businesses and individuals to have a more profound impact on the improvement of environmental sustainability (Ehrenfeld 1997).

References

Airbus ACADEMY. (2017). Airbus Corporate Answers to Disseminate Environmental Management sYstem (ACDEMY), *Eco-efficiency and Sustainability*, G6, Issues 1 http://ec.europa.eu/environment/life/project/Projects/index.cfm?fuseaction=home.showFile&rep=file&fil=ACADEMY_DesignforEnvironment.pdf. Accessed 17 July 2017.

Bernard, S. (2011). Remanufacturing. *Journal of Environmental Economics and Management, 62,* 337–351.

Bjørn, A., & Hauschild, M. Z. (2012). Absolute versus relative environmental sustainability: What can the cradle-to-cradle and eco-efficiency concepts learn from each other? *Journal of Industrial Ecology, 17*(2), 321–332.

Department for Environment. (2014). Waste legislation and regulations—Detailed guidance—GOV.UK [www document]. https://www.gov.uk/waste-legislation-and-regulations. Accessed 20 July 2017.

Despeisse, M., Ball, P. D., Evans, S., & Levers, A. (2012). Industrial ecology at factory level—A conceptual model. *Journal of Cleaner Production, 31,* 30–39.

Ehrenfeld, J. (1997). Industrial ecology: A framework for product and process design. *Journal of Cleaner Production, 5,* 87–95.

Graedel, T. E., & Allenby, B. R. (2003). *Industrial Ecology* (2nd ed.). AT&T and Prentice Hall (1st ed. 1995).

Gmelin, H., & Seuring, S. (2014). Achieving sustainable new product development by integrating product life-cycle management capabilities. *International Journal of Production and Economics, 154,* 166–177.

Gui, L., Atasu, A., Ergun, O., & Toktay, L. B. (2016). Efficient implementation of collective extended producer responsibility legislation. *Management Science, 62*(4), 1098–1123.

Guinée, J. B., Heijungs, R., Huppes, G., Zamagni, A., Masoni, P., Buonamici, R., et al. (2011). Life cycle assessment: Past, present, and future. *Environmental Science and Technology, 45*(1), 90–96.

Haynsworth, H. C., & Lyons, R. T. (1987). Remanufacturing by design, the missing link, *Production and Inventory Management,* second quarter, 24–29.

Hatcher, G. D., Ijomah, W. L., & Windmill, J. F. C. (2011). Design for remanufacture: A literature review and future research needs. *Journal of Cleaner Production, 19,* 2004–2014.

ISO 14040. (2006). *International Standard: Environmental management-life cycle assessment-principles and framework.* Geneva, Switzerland: International Organisation for Standardisation.

Jackson, S. A., Gopalakrishna-Remani, V., Mishra, R., & Napier, R. (2016). Examining the impact of design for environment and the mediating effect of quality management innovation on firm performance. *International Journal of Production Economics, 173,* 142–152.

Jayaraman, V. (2006). Production planning for closed-loop supply chains with product recovery and reuse: an analytical approach.*International Journal of Production Research, 44* (5), 981–998.

Leigh, M., & Li, X. (2015). Industrial ecology, industrial symbiosis and supply chain environmental sustainability: A case study of a large UK distributor. *Journal of Cleaner Production, 106,* 632–643.

Lifset, R. (2000). Industrial ecology: Building a framework for eco-design and life cycle assessment. *Journal of Japan Institute of Electronics Packaging, 3*(5), 403–407.

Lifset, R., Atalay, A., & Naoko, T. (2013). Extended producer responsibility. *Journal of Industrial Ecology, 17,* 162–166.

PwC. (2010). *Life cycle assessment and forest products: A white paper.* https://www.pwc.com/gx/en/forest-paper-packaging/pdf/fpac-lca-white-paper.pdf. Accessed 18 July 2017.

Silvestre, J. D., de Brito, J., & Pinheiro, M. D. (2014). Environmental impacts and benefits of the end-of-life of building materials e calculation rules, results and contribution to a "cradle to cradle" life cycle. *Journal of Cleaner Production, 66,* 37–45.

Toxopeus, M. E., de Koeijer, B. L. A., & Meij, A. G. G. H. (2015). Cradle to Cradle: Effective vision vs. efficient practice? *Procedia CIRP, 29,* 384–389.

Xiang, W., & Ming, C. (2011). Implementing extended producer responsibility: Vehicle remanufacturing in China. *Journal of Cleaner Production, 19,* 680–686.

6

Challenges for Applying Industrial Ecology and Future Development of Industrial Ecology

Abstract This chapter explores four challenges for Industrial Ecology (IE) applications: a paradigm shift from linear to closed-loop thinking, restriction lift in legislation and regulations on waste, establishment of knowledge webs, and development of symbiotic and recycling networks. Future development of IE is reflected in each of its study areas. In the area of 'industrial ecosystem', features and limitations of different types of industrial ecosystems require further exploration and extended system thinking. For IS, development of knowledge webs, symbiotic networks and infrastructure of end-life-waste collection process are further research agendas. For IM, quantification methods of resource flows in industrial ecosystems require further development. For environmental legislation and regulations, alignment with policy-makers needs to be explored in order to support IE applications on a much larger scale.

Keywords Challenges for applying Industrial Ecology (IE) and Industrial Symbiosis (IS) · Future development of Industrial Ecology (IE)

6.1 Introduction

After nearly thirty years of development, Industrial Ecology (IE) has established its position as an interdisciplinary study field for environmental sustainability development. IE is not only relevant to environmental science and management and material engineering, but also to operations management (OM) and other management studies. However, IE applications face a number of challenges, which have been classified in different ways by different researchers (Corder et al. 2014; Golev et al. 2014; Watkins et al. 2013). This chapter considers and explores four challenges.

IE contains four important study areas: industrial ecosystem, industrial symbiosis (IS), industrial metabolism (IM), environmental legislation and regulations for IE applications. This chapter presents IE future development elements across the four areas.

6.2 Challenges for Applying Industrial Ecology and Industrial Symbiosis

In the literature, some challenges or barriers to IE applications have been identified, as well as some enablers for overcoming challenges to support IE applications in practice (Bansal and Mcknight 2009; Corder et al. 2014; Geng and Cote 2004; Malcolm and Clift 2002; Yu et al. 2015). Four challenges are presented in this section, from paradigm shift and the restriction lift of legislation and regulations from a wide industrial and societal perspective to the establishment and development of knowledge webs and symbiotic and recycling networks. The first two challenges focus on IE applications in general and the other two are more concerned with IS applications to support the development of closed-loop material exchanges and energy cascading in IE.

1. *Paradigm shift from linear thinking to closed-loop thinking in industries and societies*

After nearly a decade from the first recognition of IE by Frosch and Gallopoulos (1989), IE was considered to become part of a new evolving Dominant Social Paradigm (DSP) to maintain the natural world as a fundamental normative goal' in 1997 (Ehrenfeld 1997). Twenty years later since then, IE has made some progress both in interdisciplinary development and industrial practice by embedding its principles in industries and societies. IE is no longer just a concern to environmental science, environmental management and material engineering, but also part of operations management (OM) and other management disciplines. IE has embraced and developed different approaches and analytical tools, such as design for environment (DfE), a variety of life cycle analytical tools, and cradle to cradle (C2C) (Guinée et al. 2011; Jackson et al. 2016; Silvestre et al. 2014). These approaches and tools have been increasingly applied in the product and process design to support the closed-loop industrial ecosystem development. IE applications are also reflected in IS practice through the development of many eco-industrial parks (EIPs) around the world (Gibbs and Deutz 2005; Heeres et al. 2004; Park et al. 2016; Tian et al. 2012) and a number of facilitated national IS programmes (Jensen et al. 2011; Mirata 2004). IE applications have also reached the quantification of industrial metabolism for material flows through and within EIPs to indicate the level of the closed-loop material exchanges and energy cascading within and across industrial ecosystems (Lu et al. 2015).

However, it is still hard to claim that IE has become part of a new DSP. An understanding of IE is still limited to a relatively small number of people who are engaged in environmental sustainability and pursue the practice of IE. In addition, some understanding of IE remains superficial as IE is a relatively new study field, engaging different study disciplines and complex concepts which require further development and clarification. IE has not achieved the status of a 'well-known' norm in societies and is far from being widely practised in industry, considering the total number of companies in the industrial world. It remains a challenge for IE to become part of a new DSP and organised effort is still needed. This new DSP needs to consider the closed-loop thinking in all regimes in industrial and social practices (Ehrenfeld 2004), to replace the linear transformation thinking of inputs and outputs of a process. The spread and depth of the understanding of IE and its role in developing environmental sustainability (Fig. 6.1) are far from being

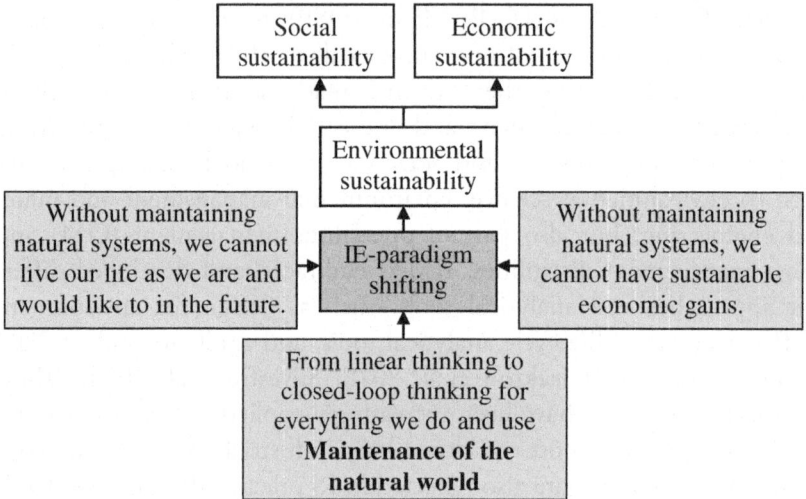

Fig. 6.1 The fundamental role of IE to environmental sustainability, and further to social and economic sustainability

well-recognised. This is further reflected by the fundamental contribution of environmental sustainability to economic and social sustainability (Goodland 1995; Goodland and Daly 1996) (Fig. 6.1). The linear process transformation is still taught and considered in OM. Product stewardship (Lewis 2005; Rogers et al. 2010) and extended producer responsibility (Lifset et al. 2013; Gui et al. 2016; Xiang and Ming 2011) largely remain as concepts and have many hurdles to overcome in practice. A total integration of different parties along and across supply chains and product life cycles to close loops of material exchanges to achieve zero disposals in the industrial world needs to happen first, but will remain a very challenging task.

A wide view of the word 'industrial' in 'industrial ecology' needs to be recognised in societies and IE is not limited in activities directly relating to industries, but includes all activities along the entire product life cycle, including product after-use. A wide perspective of 'industrial' in IE includes industrially related products and services in use and after-use, as well as industrial activities to produce goods and provide services. That proposes a challenge to the application of IE in all areas of our everyday life, industrial or societal.

2. *Restriction lift on legislation and regulations on 'waste' to support reuse, remanufacturing and recycling activities*

A number of papers emphasised that legislation and regulations on waste did not support recycling opportunities or turning waste into the feedstock to another process, which lead to IS and IE practices (Gibbs and Deutz 2005; Leigh and Li 2015; Malcolm and Clift 2002; Park et al. 2016; Watkins et al. 2013).

Legislation and regulations in relation to waste and waste management are not always straightforward in allowing companies to capture opportunities or initiatives to turn waste into resources (Leigh and Li 2015; Malcolm and Clift 2002; Park et al. 2016). The purpose of business laws for handling industrial waste and associated regulations to enforce the legislation is to ensure the success of environmental protection by restricting the disposal of hazardous waste to landfill. However, legislation and regulations on waste have been set based on an understanding of the potentially harmful effects of 'waste', if being disposed to the environment. Clearly, there is no negotiation for radiological waste, which is strictly dealt with under separate regulatory regimes (Malcolm and Clift 2002). Here the discussion refers to the legislation and regulations that deal with non-radiological waste, which can be re-entered into a production process rather than being disposed of. As IE is still a relatively novel 'idea' in industries and societies (refer to challenge 1), the long response time of a legal system generates the time lag for the consideration of the IE principle. The same applies for corresponding laws to take new ideas in developing areas into consideration.

However, the challenge for IE applications in relation to waste in legislation and regulations seems severer than necessary, as the dispute has focused on the definition and a number of interpretations of 'waste' in business law practices. The European Framework Directive defines that 'waste is any substance or object which the holder discards or intends to discard' (Malcolm and Clift 2002, p. 4). However, this has led to at least three different interpretations of waste in a number of court cases dealt with in both the European Court of Justice and the English High Court (Malcolm and Clift 2002). Whether a substance is considered as 'waste' is subject to one or more of the following three judgements:

1. Whether it is discarded or intended to be discarded,
2. Whether it requires a recovery operation before it is reusable, and/or
3. Whether it represents a burden to the holder (producer).

If yes to one of the above three questions, the substance might be classified as waste and the relevant legislation and regulations on waste are applicable. Further, what is a recovery operation needs to be clarified. In some cases, it was up to a judge to interpret the meaning of waste defined in the waste legislation in a court case, which led to inconsistent and illogical court decisions concerning industrial waste handling (detail see Malcolm and Clift 2002). Certainly, the interpretation does not always favour IE applications as IE considers that 'waste from one industrial process can serve as the raw materials for another, thereby reducing the impact of industry on the environment' (Frosch and Gallopoulos 1989, p. 144).

Critically, the issue is only partially due to multiple interpretations of waste in the legislation. It is actually a lack of understanding of the IE view towards waste and the contribution of IE to environmental sustainability. Waste need not remain waste in IE (Malcolm and Clift 2002) and there should be zero waste in the total closed-loop material exchange system - the ecosystem of the Earth. Industrial 'waste' can go through reuse, remanufacturing, recycling to re-enter into multiple product life cycles, rather than disposals. This consideration increases levels of closed-loop material exchanges as materials (or waste), depending on the perspective at different stages in the entire product life cycle, are kept in industrial (eco)systems. If the legislation and regulations reflect the IE view that allows 'waste' to flow along multiple product life cycles and among different companies without unnecessary restriction, IE and its applications can prosper. IE only allows non-harmful waste transformation and zero toxic materials entering into product life cycles to ultimately contribute to environmental sustainability of companies, supply chains, societies and beyond.

It is not the intention of these business laws and associated regulations to create obstacles for IE applications. It could be mainly because of the time lag of updating business laws, as well as the current lack of understanding of IE and its novel ways to achieve environment protection - the closed-loop material exchanges and energy cascading. IE specialists and professionals need to work closely with policy-makers to influence their thinking and understanding of the principle of IE and its fundamental role in achieving sustainability in industries and societies to reduce the time-lag effect. People working in the IE field need to improve their understanding of IE first before they are able to convince policy-makers of this novel idea for resource and waste management to overcome this legislative and regulatory challenge.

3. *Knowledge web establishment on all scales across all levels, for both technical and managerial sides*

Knowledge of potentially different exchanges from waste (industrial residues) and by-products to input resources either within the same company or across different companies and industries is fragmented in IE and IS practice. Researchers and professionals on the technical side work on technical solutions for particular types of novel exchanges, such as fly ash for making cement, pharmaceutical residues for producing fertiliser and recycling of paper or primary metals. Researchers and professionals on the management side focus on the improvement of communication and trust among companies, both of which are critical for information sharing, leading to successful synergies. A feasible technical solution alone does not guarantee its implementation and success if the managerial infrastructure and support are absent (Golev et al. 2014). Technical or managerial researchers and professionals need to have a full understanding of the skills, knowledge and responsibilities of the other side to achieve an integrated solution. Establishment of a knowledge web of possible exchanges of industrial waste (residues), by-products and energy, as well as information regarding companies' core processes which generate these waste (residues) and by-products is a challenging task. Such a knowledge web of integration impacts positively on the success of IE and IS applications and further on environmental sustainability development.

Besides, knowledge of the entire product life cycle, from material acquisition, manufacturing, distribution, use and after-use on the mainstream of a product with the coordination of the product and process design is limited in industries and academic research. IE and IS emphasise the integration of product after-use options including reuse, remanufacturing and recycling with product and process design to close loops of material exchanges. Design for environment (DfE) considers these product after-use options at the product and process design phase to enhance their effectiveness (Jackson et al. 2016). DfE has not been covered in OM and has not realised its full potential in IE and IS applications.

In addition, knowledge concerning the development of the recycling industry and its role in IE and IS applications is lacking. How the recycling industry has been developed and evolved and how it can be integrated with a variety of mainstream industries to work to its full potential still need to be explored. In addition, the involvement of consumers has been considered as a social element or part in 'urban ecology', but should also be part of IE in order to 'close' the product life cycle and contribute to zero disposals. Some research has been conducted regarding extended producer responsibility (Lifset et al. 2013; Gui et al. 2016; Xiang and Ming 2011) and product stewardship (Lewis 2005; Rogers et al. 2010), which assists the development of material- and product-based industrial ecosystems. The three types of industrial ecosystems, material-based, product-based and geographic-based with the relevance of determining boundaries of industrial ecosystems, have been addressed in Chap. 3 in considerable detail.

4. *Industrial symbiotic and recycling network development beyond local and regional geographic areas and beyond industries*

The establishment of knowledge webs of novel material, waste and energy exchanges, and associated industrial core processes (refer to challenge 3) facilitates the development of industrial symbiotic and recycling networks. These networks help companies identify, develop and establish synergies as well as help to build the infrastructure for recycling to turn waste, both industrial and consumer waste, to or recycled raw materials. The development of synergies among diverse companies, even in a planned and intended EIP poses great challenges (Gibbs and Deutz 2007). Development of the recycling infrastructure, which requires consumer involvement and is centred by the recycling industry, creates an even greater challenge.

The development of symbiotic relationships among companies requires a number of conditions, not just simply relying on technical feasibility. Apart from the restriction of the existing legislation and regulations on waste (refer to challenge 2), whether a synergy can be developed between two companies also depends on the following business operations related factors:

1. Mutually economic benefits (Branson 2016; Paquin et al. 2014),
2. Communication and trust (Golev et al. 2014; Corder et al. 2014),
3. Information sharing (Golev et al. 2014), and
4. Long-term perspective of a synergy (Mirata 2004).

A synergy needs to carry mutually economic benefits for both companies. A novel material exchange provides a solution for one company to manage its industrial waste (residues) or by-products, otherwise the waste or by-products incur a disposal cost, as well as providing the other company with an economic option for raw materials (Branson 2016). This exchange needs to be financially sound to be feasible in the long run. Communication and trust which are critical to information sharing have been emphasised in business relationship development (Golev et al. 2014) and there are no exceptions for symbiotic relationship development. These elements either act as enablers if managed well or otherwise pose barriers for companies to establish symbiotic relationships (Golev et al. 2014; Corder et al. 2014). Effective communication and trust developed between two companies lead to information sharing for developing a synergy. Sharing information from a wide perspective, such as core business processes, waste and residues, and the long-term strategic plan for novel exchanges between companies lead to successful synergies (Golev et al. 2014). As capital and human investment are often required to build an infrastructure for synergies, a long-term supply and demand relationship between companies becomes essential in order to reduce risk by committing to a symbiotic relationship (Mirata 2004). All these factors pose challenges for companies to overcome.

These conditions are not unique to symbiotic relationship development and they are also applicable to other business relationship development more widely. Greater difficulty lies in the fact that commonly symbiotic exchanges involve traditionally unrelated companies and industries. This makes initialisation, communication and trust regarding symbiotic relationships more challenging to accomplish, as this type of business relationships involve a high degree of uncertainty and risk.

The literature of IS within the study field of IE emphasises the importance of an 'organisation' acting as facilitator, either informally or formally to support symbiotic network development. A facilitating organisation can be either informal or formal. An example of an informal organisation is the social club at the early stage of the Kalundborg IS practice in Denmark (Ehrenfeld and Gertler 1997). The social club creates a platform for business executives to share their business challenges and search solutions. However, relying on a social club is a long shot for establishing business synergies and the transferability is low. Formal IS centres, like the Kalundborg Symbiosis Centre in more recent development, the Eco-Centre in the Tianjin Economic-Technological Development Area (TEDA), Tianjin, China, regional EIP centres for the nationwide EIP transformation programme in Korea and the IS coordinating centres for the UK national IS programme, host workshops and initialise synergies for a variety of companies to bypass the trust-building stage to accelerate the development of a synergy.

Waste collectors and processors in the recycling industry have an anchor role in the network of turning end-life-product waste from consumers (as well as industrial wastes from manufacturing and distribution companies) into recycled materials (Despeisse et al. 2012; Leigh and Li 2015). Different factors, which include the end-life-product waste legislation and regulations, the infrastructure for waste collection, and collaboration between recycling companies and manufacturers to create supply and demand long-term relationships, are all essential for the success of this type of network establishment and development.

6.3 Future Development of Industrial Ecology

The matrix of the four study areas in IE along with types of industrial ecosystems provides a system to present future development elements of IE for research and applications (Table 6.1). Material- and product-based industrial ecosystems share many similarities and have been considered in one category.

Some elements are shared with all three types of industrial ecosystems, such as the last two elements in the area of IM for quantifying environmental performance of an industrial ecosystem and efficiency of material and energy flows over time and space. The performance of synergies and industrial ecosystems is an area which has not been paid sufficient attention in IE studies. The reason might be due to the difficulty and complexity of data collection and accessibility, as data collection involves data of different substances, materials, waste and by-products from different parties or along the product life cycle as well as information regarding core businesses.

The development of IE in the future requires collaborative working among academics, practitioners, and governmental policy and decision makers to achieve an integrated and consistent approach for developing high levels of closed-loop industrial ecosystems. Education of IE in OM is essential at the current stage of its development, as 'business schools have been slow to pick up environment or sustainability as a legitimate area for research and teaching, …' (Ehrenfeld 2004, p. 830) including IE, and this remains the case in 2017.

6.4 Summary

This chapter has presented four challenges to IE and IS applications, including the DSP shift, legislation and regulation alignment, knowledge web establishment, and symbiotic and recycling network development. It remains a challenge for IE to become part of a new DSP to influence individual and business thinking and decisions. Recognition of the fundamental role of IE to environmental sustainability, which

Table 6.1 Future development elements of IE in four areas for research and applications

	Geographic-based industrial ecosystems	Material- and product-based industrial ecosystems
Industrial ecosystem	• Identify features of geographic-based industrial ecosystems • Explore limitations of geographic-based industrial ecosystems • Apply extended system thinking	• Identify features of material- or product-based industrial ecosystems • Explore limitation of material- or product-based industrial ecosystems • Improve transformation of a material from one product to another efficiently
Industrial Symbiosis (IS)	• Develop knowledge webs of material exchanges and energy cascading • Develop symbiotic networks of companies locally, regionally and globally	• Integrate different material- and product- based industrial ecosystems through industrial collaborations • Develop the recycling industry • Develop the infrastructure for end-life-product waste collection and processing
Industrial Metabolism (IM)	• Quantify the rate of each industrial collaboration in terms of material exchanges and energy cascading • Quantify the percentage of internal material and energy flows compared to the total material and energy flows • Quantify environmental performance of an industrial ecosystem • Quantity efficiency of material and energy flows over time and space	• Quantify the status of an industrial ecosystem in terms of the rate of its materials transformation and energy cascading • Quantify the level of the closed-loop within a product life cycle • Quantify environmental performance of an industrial ecosystem • Quantity efficiency of material and energy flows over time and space
Legislation and regulations for IE	• Study industrial waste legislation and regulations required for synergies • Identify issues in the existing legislation and regulations on waste and waste management	• Study end-of-product waste legislation and regulations required to support recycling of consumer waste • Identify issues in the existing legislation and regulations on end-of-waste recycling

supports economic and social sustainability, is yet to come. This paradigm shift requires culture change in industries and societies in order for IE to be applied on a much larger scale to fulfil its full potential for environmental sustainability development.

The current legislation and regulations on waste and waste management also present a challenge for IE and IS applications. The understanding of using waste from one process as resources for another is critical for IE principles to be incorporated into legislation and regulations to remove legal barriers for companies to apply IE and IS.

Without well-established knowledge webs that integrate technical solutions and managerial aspects, IE and IS applications remain fragmented and the scale of the transformation of industrial systems to industrial ecosystems is restricted. Symbiotic network establishment requires planned IS programmes, which are facilitated and supported by IS coordinating centres and teams. Trust building leading to information sharing needs to be further enhanced by the continuous evolution of IS networks on a local, regional, national and international scale, throughout industries and societies.

These challenges are not isolated and resolving one challenge may lead to some degree of resolution of others. For example, alignment of legislation and regulations on waste with IE principles and establishment of knowledge webs to facilitate the development of symbiotic and recycling networks contribute to a paradigm shift of IE to be part of the new DSP.

Further development of IE is considered in its four study areas. In the area of industrial ecosystem, extended system thinking for the development of different types of industrial ecosystems and the identification of features and limitations of each type of industrial ecosystem require further exploration. In IS, ways to establish a knowledge web and symbiotic networks on all scales across all levels are yet to be developed systematically. In IM, methods for quantifying levels of material, energy and waste flows and their efficiency require further development. In legislation and regulations for IE, approaches to align legislation and regulations with the principle of IE need to be integrated among industries and governmental regulators. The future of IE is full of promise, but also challenges. As the need for improving environmental sustainability is constantly increasing, the essential role of IE and other ecologies, such as urban ecology, would be appreciated and even demanded far more in societies and industries. Integration of different types of ecology as well as with biological ecology will form a greater driving force to move our societies and industries to much higher levels of closed-loop material exchanges and efficiency of energy cascading, leading to an advanced state of sustainability overall and a closed-loop society (Ehrenfeld 2004). Failure to do so will inevitably result in global and catastrophic consequences to human populations and never mind the maintenance of current living standards for future generations to come.

References

Bansal, P., & McKnight, B. (2009). Looking forward, pushing back and peering sideways: Analyzing the sustainability of industrial symbiosis. *Journal of Supply Chain Management, 45,* 26–37.

Branson, R. (2016). Re-structuring Kalundborg: The reality of bilateral symbiosis and other insights. *Journal of Cleaner Production, 112,* 4344–4352.

Corder, G. D., Golev, A., Fyfe, J., & King, S. (2014). The status of industrial ecology in Australia: Barriers and enablers. *Resources, 3,* 340–361.

Despeisse, M., Ball, P. D., Evans, S., & Levers, A. (2012). Industrial ecology at factory level—A conceptual model. *Journal of Cleaner Production, 31,* 30–39.

Ehrenfeld, J. (1997). Industrial ecology: A framework for product and process design. *Journal of Cleaner Production, 5,* 87–95.

Ehrenfeld, J. (2004). Industrial ecology: A new field or only a metaphor? *Journal of Cleaner Production, 12,* 825–831.

Ehrenfeld, J., & Gertler. (1997). Industrial ecology in practice: The evolution of interdependence at Kalundborg. *Journal of Industrial Ecology, 1,* 67–79.

Frosch, R. A., & Gallopoulos, N. E. (1989). Strategies for manufacturing. *Scientific American, 261*(September), 144–152.

Geng, Y., & Côte, R. (2004). Applying industrial ecology in rapidly industrialised Asian countries. *International Journal of Sustainable Development and World Ecology, 11,* 69–85.

Gibbs, D., & Deutz, P. (2005). Implementing industrial ecology? Planning for eco-industrial parks in the USA. *Geoforum, 36,* 452–464.

Gibbs, D., & Deutz, P. (2007). Reflections on implementing industrial ecology through eco-industrial park development. *Journal of Cleaner Production, 15* (17), 1683–1695.

Golev, A., Corder, G. D., & Giurco, d. P. (2014). Barriers to Industrial Symbiosis: Insights from the use of a maturity grid. *Journal of Industrial Ecology, 19,* 141–153.

Goodland, R. (1995). The concept of environmental sustainability. *Annual Review of Ecology and Systematics, 26,* 1–24.

Goodland, R., & Daly, H. (1996). Environmental sustainability: Universal and non-negotiable. *Ecological Applications, 6,* 1002–1017.

Gui, L., Atasu, A., Ergun, O., & Toktay, L. B. (2016). Efficient implementation of collective extended producer responsibility legislation. *Management Science, 62,* 1098–1123.

Guinée, J. B., Heijungs, R., Huppes, G., Zamagni, A., Masoni, P., Buonamici, R., et al. (2011). Life cycle assessment: Past, present, and future. *Environmental Science and Technology, 45,* 90–96.

Heeres, R. R., Vermulen, W. J. V., & de Walle, F. B. (2004). Eco-industrial park initiatives in the USA and the Netherlands: First lessons. *Journal of Cleaner Production, 12,* 985–995.

Jackson, S. A., Gopalakrishna-Remani, V., Mishra, R., & Napier, R. (2016). Examining the impact of design for environment and the mediating effect of quality management innovation on firm performance. *International Journal of Production Economics, 173,* 142–152.

Jensen, P. D., Basson, L., Hellawell, E., Bailey, M. R., & Leach, M. (2011). Quantifying 'geographic proximity': Experiences from United Kingdom's National Industrial Symbiosis Programme, Resources. *Conservation and Recycling, 55,* 703–712.

Leigh, M., & Li, X. (2015). Industrial ecology, industrial symbiosis and supply chain environmental sustainability: A case study of a large UK distributor. *Journal of Cleaner Production, 106,* 632–643.

Lewis, H. (2005). Defining product stewardship and sustainability in the Australian packaging industry. *Environmental Science & Policy, 8,* 45–55.

Lifset, R., Atalay, A., & Naoko, T. (2013). Extended producer responsibility. *Journal of Industrial Ecology, 17,* 162–166.

Lu, Y., Chen, B., Feng, K., & Hubacek, K. (2015). Ecological network analysis for carbon metabolism of eco-industrial parks: A case study of a typical

eco-industrial park in Beijing. *Environmental Science and Technology, 49,* 7254–7264.
Malcolm, R., & Clift, R. (2002). Barriers to Industrial Ecology, The strange case of "The Tombesi Bypass". *Journal of Industrial Ecology, 6,* 4–7.
Mirata, M. (2004). Experiences from early stages of a national industrial symbiosis programme in the UK: Determinants and coordination challenges. *Journal of Cleaner Production, 12,* 967–983.
Paquin, R. L., Tilleman, S. G., & Howard-Grenville, J. (2014). Is there cash in that trash? *Journal of Industrial Ecology, 8,* 268–279.
Park, J. M., Park, J. Y., & Park, H.-S. (2016). A review of the eco-industrial park development program in Korea: Progress and achievement in the first phase, 2005-2010. *Journal of Cleaner Production, 114,* 33–44.
Rogers, D. S., Rogers, Z. S., & Lembke, R. (2010). Creating value through product stewardship and take back. *Sustainability Accounting, Management and Policy Journal, 1,* 133–160.
Silvestre, J. D., de Brito, J., & Pinheiro, M. D. (2014). Environmental impacts and benefits of the end-of-life of building materials—Calculation rules, results and contribution to a "cradle to cradle" life cycle. *Journal of Cleaner Production, 66,* 37–45.
Tian, J., Shi, H., Chen, Y., & Chen, L. (2012). Assessment of industrial metabolisms of sulfur in a Chinese fine chemical industrial park. *Journal of Cleaner Production, 32,* 262–272.
Watkins, G., Husgafvel, R., Pajunen, N., Dhl, O., & Heiskanen, K. (2013). Overcoming institutional barriers in the development of novel process industry residue based symbiosis products—Case study at the EU level. *Minerals Engineering, 41,* 31–40.
Xiang, W., & Ming, C. (2011). Implementing extended producer responsibility: Vehicle remanufacturing in China. *Journal of Cleaner Production, 19,* 680–686.
Yu, C., Dijkema, G. P. J., & Jong, M. D. (2015). What Makes Eco-Transformation of Industrial Parks Take Off in China? *Journal of Industrial Ecology, 19* (3), 441–456.

Index

A
After-use product. *See* product after use

B
Balance
  production and decomposition 27
Biological ecosystem 3, 6, 25, 27, 34, 35
Boundary
  extended system 29, 66
By-product 2, 45, 46, 52, 55, 63, 67, 68, 71, 81, 117

C
Carrying capacity 3, 6, 11
Closed-loop
  industrial ecosystem 11, 19, 25, 27, 30, 34, 35, 42, 58, 67, 92, 98, 102, 113, 121

  material exchange 5, 7, 11, 25, 27, 29, 33, 40–42, 47, 49, 53, 55, 62, 67, 80, 92, 93, 96, 98, 102, 112, 113, 116, 117, 124
  representation 2
  system 3–6, 11
  thinking 35, 107, 113
Collaboration
  cross-boundary 84
  symbiotic 58, 63, 64, 66, 69, 72, 75, 78, 77
Cradle to cradle (C2C) 95, 113
Cradle to grave (C2G) 96

D
Dalian Economic-Technological Development Zone (DETDZ) 77
Degradability 97
Dematerialisation 41

Design for
 disassembly 98
 environment (DfE) 4, 41, 96, 100, 113, 118
 multiple lifecycle 100
 recycling (DfRec) 98, 100
 remanufacture (DfRem) 98, 99
 remanufacture process 98
 reuse 96, 99
 upgrading 100
Diverse industrial involvement 78
Dominant Social Paradigm (DSP) 25, 36, 107, 113

E

Eco-design options 96, 102
Eco-efficiency 33, 51, 52
Eco-industrial cluster (EIC) 83, 85, 86
Eco-industrial park (EIP) 61, 62
 demonstration EIP 77
 pilot EIP 77
Ecosystem of the Earth 6, 7, 11, 27, 34, 36, 42, 58, 116
Energy cascading
 efficiency of 5, 35, 47–49, 53, 62, 80, 92, 93, 96, 98, 102, 103, 106, 124
Environmental sustainability 2, 3, 7, 10, 36, 41, 48, 54, 55, 58, 69, 70, 82, 96, 107, 112–114, 116, 117, 121, 123, 124
Extended producer responsibility (EPR) 45, 114, 118
Extended system thinking(view) 42, 69, 88, 123

F

Feedstock 4, 5, 46, 54, 72, 105, 107, 115

G

Geographic proximity 10, 29, 31, 62, 71, 80–83, 86

I

Industrial cluster 78, 80, 83–86
Industrial Ecology (IE)
 applications 39, 40, 41, 57, 112, 113, 115, 116
 definitions 11
 framework 90, 100, 101
Industrial ecosystem
 boundary 29, 40, 41, 47, 55, 86
 geographic-based 42, 46, 55, 80, 86
 material-based 42, 43, 45, 55, 118
 nearly closed-loop 6, 11, 19, 35, 80, 107
 product-based 39, 42, 45, 52, 102, 118
Industrial Metabolism
 rate 41, 113
Industrial Symbiosis (IS)
 application 7, 10, 34, 49, 50, 68, 73, 81, 123
 coordinating centre 61, 120, 123
 definition 70
 planned or facilitated 30, 50, 71, 72
 self-organised, self-organising 48
Integration 39, 91, 100, 102, 114, 117, 118, 124

Interdisciplinary
  development 113
  study field 2, 11, 25, 34, 112
  study topics 112

K
Kalundborg
  IS community 48, 57, 63
  symbiosis centre 63, 64
Knowledge webs 34, 35, 47, 48, 57, 62, 118, 123

L
Legislation and regulations on waste 39, 53, 57
Life cycle
  analysis 92–96, 106
  analytical tools 94, 95, 106, 113
  assessment (LCA) 94, 95, 106
  cost analysis (LCC/LCCA) 94
  economic analysis (LCEcA) 95
  environmental analysis (LCEnA) 95, 106
  impact assessment (LCIA) 94
  inventory (LCI) 94
  risk analysis (LCRA) 94, 95
  social analysis (LCSoA) 94, 95, 106
  sustainability analysis (LCSA) 95, 106
  thinking 7, 92–96, 106
Linear transformation process 2–4, 7
Linear transformation thinking 1, 3, 113

M
Material, energy and waste exchange 9, 35, 47, 62, 104

Material
  non-recycled 52
  recycled 4, 5, 51, 97, 100, 103, 105, 107, 118, 120

N
National IS programme (NISP) 7, 30, 48, 49, 61, 62, 69, 113, 120
Natural ecosystem 13, 16, 34, 42, 102

O
Original equipment manufacturer (OEM) 98
Overall industrial ecosystem 12, 82

P
Preventative integrated end-of-life approach 98
Product
  after-use 42, 44, 53, 54, 57, 95, 98, 105, 106, 114, 118
  after-use option 3, 93, 95, 102
  eco-design option 89, 94
  life cycle 42, 45, 54, 57, 91–97, 103, 105, 106, 114, 116–118, 121, 122
  life cycle analytical tool 91, 93, 94
  reconditioning 98, 99
  remanufacturing 45, 57, 93
  stewardship 97, 118

R
Recyclability 44, 52, 95, 97
Recycling
  -after-design for recycling 97, 99
  company 43, 45, 69

industry 43, 46, 54, 57, 99, 104, 118–120, 122
Regional community IS practice 48, 61, 62, 64
Remanufacturing 118
  -after-design for remanufacturing 43, 97, 99
  Remanufacturability 97
Reuse
  -after-design for reuse at the product/unit level 98, 99

S

Symbiotic
  collaboration 20, 63–65, 68, 76
  exchange 20, 67, 68, 71, 76, 81, 82, 120
  relationship 16, 17, 19, 29, 30, 33–35, 40, 41, 54, 57, 58, 76, 81, 83, 119, 120
Symbiotic relationships 48
Symbolic
  collaboration 66
  relationship 66

Synergy
  cross-boundary 82

T

TEDA 77, 81
Totality 27, 29, 30
Transferability (transferable) 66, 120
Transferability 48, 57
Transformation 1, 2, 6, 7, 11, 14, 16, 35, 36, 43–45, 48, 49, 51, 52, 55, 61, 77–80, 85, 86, 95, 114, 116, 120, 122, 123

W

Waste collector and processor 45, 104, 107, 120
Waste
  management 54, 55, 70, 115, 117, 122, 123
  non-recyclable 51, 57
  recyclable 46, 51, 52

Z

Zero disposal 54, 114, 118

The manufacturer's authorised representative in the EU is Springer Nature Customer Service Centre GmbH, Europaplatz 3, 69115 Heidelberg, Germany. If you have any concerns regarding our products, please contact ProductSafety@springernature.com

Printed and bound by CPI Group (UK) Ltd, Croydon, CR0 4YY
23/03/2026
02076402-0005